PRAISE FOR
BEYOND HUMAN

"*Beyond Human* offers a critical but enthusiastic view from a physics and biological perspective. . . . Interviews with pioneers and participants in enhancement research, robotics, and engineering, and informed perspectives on the theory, economics, and actuality of life extension, give *Beyond Human* the flavor of a tourist guide to the future, conducted by natives, rather than a missionary's critical, hands-off examination. . . . Serve[s] up a feast with many side dishes."
—*Nature*

"For those who don't follow cybernetics, robotics, and nanotech closely, Benford and Malartre offer a trenchant summary of current developments and future possibilities of how these technologies may impact human biotechnological 'self-improvement' and how those improvements may affect human societies."
—L. E. Modesitt, Jr.

"A fascinating and plausible look at the future direction of human society. . . . Nothing in the book seems far-fetched or implausible; everything is solidly based in current research (much of which, like the bit about artificial tongues, will astonish readers unfamiliar with cutting-edge technology). An engagingly written, eye-opening, and well-documented book."
—*Booklist*

"Nontechnical and thus well suited to general readers. At the same time, [the authors] cite examples galore from SF to illustrate their points, which should make the book of extra interest to readers with SF backgrounds."
—*Analog*

"Beyond speculation, the book cites numerous present-day examples of this trend toward everyday robotics and technology-enhanced humans. This brave new world is already here, the authors argue."
—*Science News*

Other Nonfiction Books
by Gregory Benford

Deep Time: How Humanity Communicates Across Millennia

Beyond Human

Living with Robots and Cyborgs

Gregory Benford

and

Elisabeth Malartre

A TOM DOHERTY ASSOCIATES BOOK

NEW YORK

BEYOND HUMAN

Copyright © 2007 by Gregory Benford and Elisabeth Malartre

Book design by Mary A. Wirth

A Forge Book
Published by Tom Doherty Associates, LLC
175 Fifth Avenue
New York, NY 10010

www.tor-forge.com

Forge® is a registered trademark of Tom Doherty Associates, LLC.

Library of Congress Cataloging-in-Publication Data

Benford, Gregory.
    Beyond human : living with robots and cyborgs / Gregory Benford and
    Elisabeth Malartre.
        p. cm.
    Includes bibliographical references.
    ISBN-13: 978-0-7653-1083-5
    ISBN-10: 0-7653-1083-X
    1. Robots. I. Malartre, Elisabeth. II. Title.

TJ211.B465 2007
303.48'34—dc22

                                                    2007018360

First Hardcover Edition: September 2007
First Trade Paperback Edition: December 2008

Printed in the United States of America

P1

Contents

Part II: Robots Plus

Part III: 'Bots, 'Borgs, Bionics, and Betters

Preface

The next decade promises another qualitative shift in the way we view technology, when the ideas treated here condense into firm, available objects—robots, cyborg parts, and the many variations in between. These innovations will alter our lives and the way we view our world as much as, and probably more than, the advent of personal computers during the 1980s.

The coming of the robots will be obvious. They will first work inside buildings with well-defined geometries, carrying paperwork or packages. Then they will be security guards, prowling company corridors through the night, using infrared vision in dark areas. We will leave a robot house sitter behind when we go on trips, checking in over the Internet to literally see what's going on. It will roll through the house it knows from experience, noting any changes and sounding an alarm if these alterations appear threatening. With an arm/hand combination that can open doors or turn off the oven in case you left it on, and infrared eyes that see in the dark, it can be more versatile than a neighborhood security guard.

Soon robots will be everywhere, performing surgery, exploring hazardous places, making rescues, fighting fires, handling heavy goods. After a decade or two, they will be as unremarkable as the computer screen is now in offices, airports, and restaurants. Each new advance will create a momentary flurry of excitement, but the robots will increasingly blend in. Already, we see the term used even for software, as with a Japanese Web site that promises, "An e-mail from our robot will be delivered to you for confirmation."

The cyborgs will be less obvious. Many changes will be hidden from view.

At first these additions to the human body will be interior, as rebuilt joints, elbows, and hearts are now. Then larger adjuncts will appear, perhaps on people's heads or limbs. Soon we will cross the line between repair and augmentation, probably first in sports medicine and the military, then spreading to everyone who wants to make their body perform better than it ordinarily could. Physical performance will come first, but mental agility will follow, for the same reason that we now drink coffee before starting work. Controversy will arise, with many saying we are assuming more power than people should have over themselves. But such sentiments will not stop the desire to be better than we are, or to extend our active life spans; they never have.

This book treats the landscape of human self-change and robotic development as poles of the same general phenomenon. It is an ever-shifting terrain. To capture it, we have incorporated interviews with workers and thinkers in these fields. We wanted to reproduce, as often as possible in their own words, expert views of what may lie ahead.

Technology is moving so fast and unpredictably that some bits of what we have written will be out of date by the time you read this. Some fields will surge ahead while others will simply fade away, and new, more promising approaches will emerge. But the broader implications of imminent change will remain.

Science has often followed cultural anticipation, not led it. Fiction and film have meditated upon the upcoming social issues of robots and cyborgs for centuries. From the Jewish folktale of the golem and Thomas Edison's attempts to build an artificial woman (as described in *Edison's Eve*) to myriad science fictions, our culture is struggling with these ideas.

This is not really new. Much technology, and even science, is connected firmly with its social context, and has sometimes even arisen from it. For example, in 1932 physicist Leó Szilárd read H. G. Wells's 1914 novel, *The World Set Free,* which predicted the discovery of artificial radioactivity in 1933—a direct hit, as it turned out. The novel depicted atomic power, bombs, and a world war between an alliance of England, France, and perhaps the United States, against Germany and Austria. Wells's fictional bombs probably began the misnomer "atomic" instead of "nuclear," but they did work by "chain reactions." The novel was dedicated to Frederick Soddy, whose study of radium gave Wells the idea. Szilárd saw the possibility of such weapons, and for decades was a central driver toward first making and then controlling them. He got Einstein's signature on the letter to President Roosevelt that launched the Manhattan Project.

Many cultures have imagined altering ourselves and duplicating human abilities in machines. Much modern science fiction has imagined and thought through the personal and social effects of doing this, in well-realized, realistic narratives. Such work can be a valuable guide to navigating the ever-rushing waters of change in the next century.

In this book we note how often research in human augmentation and robotics has stemmed from speculations and projections in the surrounding culture. Writers ventured to imagine robots long before scientists—who had read those earlier works, often in adolescence—tried to make them come true. Such ideas interweave with the concrete, technical advances treated in a cultural dance that we wish to illuminate.

Our long voyage toward artificial enhancement and even replacement of ourselves is centuries old, but just beginning. Here we attempt a snapshot of that journey, examining the current interface between biology and electronic technology. We do so knowing that within a few years many predictions we make will probably prove to be either far out at sea, or left high and dry.

i

Man Plus

Augmentation:
From an Old Tradition to
Our Brave New Bionic World

The point of technology is to extend what we can do with our bodies,
our senses, and, most of all, our minds.

—*Scientific American*

The urge to improve upon the human body is at least as old as our species. It dates back to the first time a protohuman picked up a rock to smash a bone to reach the hidden marrow, or used a stick to help dig down to a tender root.

Compared with other animals, we humans don't seem very well prepared to defend ourselves or to wrest a living from the wilderness. We don't see or hear as well as other hunting animals. We don't have claws for digging or sharp teeth for ripping meat. We have little fur for protection: Our skin is thin and tender. We have soft feet and can't run as fast as many other animals, be they prey, predator, or competition.

So what was the edge that allowed us not only to survive but to dominate the planet? We used other objects. We now call it technology, and treat our tools and machines as though they are separate from us. But they are and have always been simply augmentations of our bodies. They make us stronger, faster, and they restore hearing and vision.

What was the first tool used by humans or their ancestors? Without a time machine, we will probably never know. Chimpanzees and some birds use sticks to probe for insects, and sea otters smash shells of abalone with stones gathered for that purpose. Human ancestors used stone tools at least as long as a million years ago. The use of wooden spears and digging sticks is certainly very old, but wood doesn't preserve as well as stone tools, so the direct evidence is gone. We have, however, found the stone tools that sharpened the wood. By extending the length of the human arm, spear throwers made of sticks enabled early hunters to hurl sharpened spears harder and farther. When stone shaping became sophisticated, humans created small sharp points, mounted them on thin shafts, and invented arrows.

To survive in a new environment, humans alter their technology, not themselves. Culture can change radically faster than genetics. Cold weather demanded shelters. Evidence of tent-peg holes dating back fifty thousand years, to the last ice age, has been found in France at a Neanderthal site.

Although we don't think of them as tools, clothes are a critical augmentation. Lacking built-in pouches, our ancestors had to use animal-skin slings, woven bags, or other objects to carry food, and later on, tools. More substantial clothing enabled our ancestors to survive cold spells and colonize even arctic areas, although needles for sewing were not used until twenty-six thousand years ago.

One of the most far-reaching technological innovations was the harnessing of energy, starting around 250,000 years ago. From simple wood campfires we progressed to fossil fuels through peat, soft coal, anthracite, petroleum, and natural gas. Ultimately we unlocked a source as old as the earth—nuclear fission of radioactive elements in the Earth's crust.

Tapping these energy resources enabled us to greatly extend our control over the biosphere. If we seek one characteristic to define human uniqueness, our use of external energy is a prime candidate. Many animals and plants use the sun to warm themselves, but no other animal intentionally creates fire (although a few have learned how to benefit from it).

Still, until James Watt improved the steam engine and started the Industrial Revolution two hundred years ago, humankind was stuck in a relatively low-energy niche, relying mostly on draft animals to power agriculture. Once we broke out of that subsistence mode, technology rapidly increased in power and sophistication. The application of electricity, a mere lab curiosity for a hundred years, to the telegraph signaled the start of the communications revolution we are in today.

Throughout our long journey from the African veldt, our use of technology has enabled us to colonize so many different environments on the planet that we now inhabit all but the deep ocean floors.

Our most recent technology is electronic. With a momentum largely unforeseen even in science fiction, electromagnetism went from an abstract theory to concrete technologies in a few decades. People experienced this through an ever-faster parade of marvelous devices: the telegraph, Edison's phonograph, then Marconi's radio, telephones, followed by television and onward to the myriad ways of communicating and entertaining ourselves. Beneath these were more subtle technologies. Cars and airplanes could not move without electrical firing systems to drive pistons and control movement. When computers broke over human culture in an increasingly powerful wave in the last half of the twentieth century, no one foresaw the advent of small, powerful laptops, the Internet, and the culture these innovations spawned. That wave is still sweeping away traditional methods of doing business, and even redefining our notions of society by erasing boundaries imposed by geography.

The Internet holds the promise of linking all of humanity into a giant interactive community. Some have likened it to a superorganism. Certainly computing power on silicon chips is transforming how we get information from and manipulate the world as individuals.

The concept of digital life, not rooted in flesh, has spawned books and movies about "uploading" human consciousness into computers, or "downloading" them into metal bodies. It makes for great entertainment, but when it comes down to ourselves, most of us are going to be very careful about the choices we make. The first question any seer into the future should ask is not what is technically *possible,* but rather, what is most *desirable.* In our market economy, demand drives supply.

For cyborgs particularly, two common assumptions seem dubious: first, that people will readily take to invasive technology; second, that wedding the human brain to computer assistance will be simple, and will come soon. To elicit why these appear unlikely, we asked biologist Gregory Stock, the director of the Program on Medicine, Technology, and Society at the University of California, Los Angeles. He has thought deeply about the relationship between humans and their technology. His most recent relevant book is *Metaman: The Merging of Humans and Machines into a Global Superorganism.*

Stock feels that the miniaturization of computers is about to pass a fresh threshold: When intelligence and sophistication can be embodied in devices that are either readily portable or even implantable, forming an intimate connection with our own bodies, they become sidekicks, companions.

INTERVIEW WITH GREGORY STOCK

Q: And what do these devices promise—longevity, comfort, efficiency?

A: These biological manipulations are technological changes on a substrate of biology. At the same time the technology that we see as hard technology, like computer chips, is beginning to achieve a complexity, a richness and function that make it almost biological, almost lifelike. So the sharp line between biology and nonbiology is really fading.

Q: You mean nonbiological machines performing biological functions?

A: The dividing line between harder technologies, using silicon and metals, and the biological substrates of carbon and nitrogen and water, is fading. Hard technology is achieving a lot of complexity, to do things that previously could only be done by biological entities.

Q: Such as precision measurement coupled with expert systems software, which can already diagnose many medical conditions, for example?

A: Yes. This opens up enormous possibilities, especially because the rate of evolution of nonbiological devices is so rapid. Most people believe that Moore's law—which is of course not really a law, but an observation that computing power roughly doubles every couple of years—will continue for at least fifteen years and perhaps much longer before it slows. Really enormous functionality and very, very cheap prices, with applications we are only beginning to understand.

Q: What can computing technology do for us?

A: I think that the most powerful impacts on us will *not* be to actually replace normal functionality to enhance ourselves. We are not going to put chips in our brains like you see in Hollywood with *Johnny Mnemonic,* where we can download information into a chip that is somehow flawlessly, seamlessly merged with biology. I think that is very, very distant—not because the capacity of computers will not become powerful, or be equally powerful to that of the human mind, for instance—but because the interface will be very, very messy and difficult.

I see coming replacement, even internally, to *supplement* functionality that has been lost in humans. For instance, for a person who is deaf, we have an implant that will restore some hearing. Or a pacemaker that helps a heart to function normally or effectively for a much longer period of time.

Q: Beyond the relatively few people who have lost their sight or hearing, will it be bigger than that?

A: When enthusiasts for technology imagine things that really transform human life, they tend to think of enhancements because you know we can all relate to that. Restoration of degraded function in some way—an ocular implant or a pacemaker—only affects a few people and isn't the image of what human potentials are really all about.

I personally believe that actual enhancements will not occur through implantation, because the interface with our biology is so messy. If you are functioning normally the risk versus reward structure, very much argues against it. You are not going to go through a brain surgery in order to get a little bit of additional memory, especially when you can get it through an intimate external interface with machines.

So I see the things that will affect who we are as human beings are going to be very tight, even intimate connections with machines, but *not* ones that are implanted.

As an example, we may have eyeglasses that enhance the reality around us, that merge artificial reality in by layering information on top of the background of normal reality. Labeling things—putting names to people's faces, supplementing them, recoloring them. A man coming home from work sees a storefront, and layered over it is the shopping list sent over by his wife. Things of that sort.

Beyond that, there are the transformations that are going to affect who we are biologically. The combination of technology and increasing genetic knowledge is going to make it possible to do biological interventions that will substantially modify us. For instance, it would change everything about the way we live if our life spans were doubled, if we could retard aging; and it is becoming very plausible because already there have been genetic interventions in simpler creatures—fruit flies and nematodes, producing a doubling of life span.

Any intervention like this in humans would be very multifaceted. There would be many genes involved, but it is certainly quite plausible that we would be able to retard aging or delay the onset of the degenerative process, and this would change the way we live and see ourselves—the whole trajectory of human life.

Q : Can we do that efficiently through nonbiological means?

A : We are going to always choose the least intrusive—the simplest, most reversible intervention that is possible. If you can take a drug and get performance enhancement, that's what you'll do.

You are not going to want to mess around with your genetics or implant something that is unpleasant and problematic.

So it will always be the simplest way. First pharmacology, then any sort of external technological device that we can have, later implanted perhaps, but only after it is very mature.

Technology is moving so rapidly that you wouldn't want to implant anything in the body that is going to seem archaic in a few years. You are not going to go dig back in and repeat a surgery. That is definitely something people will want to avoid. Usually in science fiction movies the wound healing and issues like that are either ignored, or you pass a force field over the body or something like that, and it heals instantly. But life isn't really like that, and people want to avoid these sorts of things.

By advancing beyond our current forms, we can better look back and define ourselves. All the terms used to describe the coming forms that will go beyond the human norm—androids, cyborgs, bionic people, robots—summon up the question of what it means, deep down, to be human.

Is it athletic prowess? (Probably not fundamental, but often depicted in films because it is easy to show.)

Intellect? (But computers press our abilities in many areas already.)

Creativity? (How creative is chess? Yet the world champion is now a computer.)

Ever since the nineteenth-century song "John Henry," we have given grudging ground to machine capabilities. (He was a steel-drivin' man, and he beat that steam drill. . . .) But after beating the new rail-laying machine, he fell dead of exhaustion.

A still greater anxiety lies below all these terms: the description of human parts as machinelike, and replacing them with actual machines, summons up the question of how much we are already essentially mechanical, right down to our minds. Hovering behind this nervousness lie questions of free will and just how much we truly control ourselves. This weds with Freud's revelations about unglimpsed impulses emerging from our unconscious. Who wants to go *there*?

Some do—if only to make a buck. In myriad science fiction films, autonomous humanlike beings mimic all manner of human responses to the world. Some get away with it.

In the 1984 film *The Terminator*, a killer robot gets more adept at imitating humans as it moves through Los Angeles. When someone addresses it, a pull-down menu of possible responses appears in its field of view, ranging from swear words to polite evasions. It learns to imitate voices from a single sentence. It can fool others over the telephone, even getting right the tones of a mother's concern so that her daughter suspects nothing (a truly terrifying scene, quietly delivered).

Such adaptability and skill at seeming human, while not being truly so, calls up a fundamental fear that we will not be able to use our ancient primate skills of sight, sound, and smell to detect machine deceptions. In the film, dogs can smell killer robots, but not after the robots learn to grow human skin over their metal bodies. This seems unlikely, since skin allows telltale scents, such as oil, to filter through it. But as is often true in films, this fact has symbolic truth: seeming lifelike enough is the essence.

Similarly, the robotic women in the original film *The Stepford Wives* still seem sexy to men who know full well they are fake. This raises the uneasy question of whether we will be willing to go along with machines pretending to be human, if they satisfy our hungers and are reasonably adept. Some of us will simply not care whether others are robots or cyborgs, and will treat them all as objects to be used.

Analyzing ourselves as machines quickly calls forth those who argue that something mystically human transcends our physical basis. This blends into the concept of "emergent phenomena"—recognizing that very complex behavior can come from simple rules. The seemingly infinite complexity of human culture could emerge, this suggests, from fairly simple patterns of logic, laid down long ago in the human mind and body by evolutionary pressures.

Increasingly, science fiction writers have spoken for the widespread, gut-level stresses we feel. Androids as sexual partners appeared in such titles as *The Silver Metal Lover* (Tanith Lee, 1982) and *The Hormone Jungle* (Robert Reed, 1988). The literature resounds with the earlier themes of robot revolution, machine takeover of society, and duplication of people without anyone knowing. These concepts make for simple plot structures but seldom explore the depth that coming technology seems likely to force us to face.

Even so influential a film as *Blade Runner* (1982) expressed profound anxiety about "androids" who were in fact completely fabricated humans, apparently made of organic parts, though with short life spans. They are termed "replicants"—as if they were mere replicas of us. This future rain-drenched Los Angeles features artificial animals (because the real ones are extinct) and many sickly humans (because most of humanity has left the polluted planet). Android replicants can be spotted with an empathy test, because they don't have any "real" feelings toward the natural world; they aren't part of it, after all.

The mere fact that these terms are routinely scrambled—robots, 'droids, replicants, bionics—and blithely misused in the popular culture tells us something also. *All* artificial forms are suspect.

Clones seem to many also suspect. These we treat later, and of course they are artificial only in a genetic sense.

The spectrum of the unnatural goes very deep. People assembled from cadaver parts are certainly not to be trusted. Moviegoers have consistently called the most famous monster "Frankenstein," but in Mary Shelley's 1818 novel, *Frankenstein: or, the Modern Prometheus,* it is the name of the creator. This most famous of all modern humanlike creations, fashioned from dead body parts, is the work of horrific overambition. It expresses the fear that our creations can put us in their dark, looming shadow.

The enormous cultural legacy of *Frankenstein* speaks to us now, nearly two centuries since a young woman created those images and ideas, of how much we distrust any attempt to ape or surpass us. The tradition includes even Thomas Alva Edison's attempts to make an artificial woman, emerging perhaps, from a basic anxiety.

Can we go too far in making ourselves machine-like, or making machines resemble us? Many feel so already. And once made, will such creatures be like us, but end up not *liking* us? These questions will arise in myriad ways in the next few decades, as we press today against boundaries that only a short while ago we met only in works of the imagination.

You, Too, Can Be Superhuman

How close is science to making us
stronger, tougher, smarter, just plain better?

Humans are general-purpose animals—perhaps the best.

Our bodies aren't best at any one sense or strategy, but we do a lot of things well. Dogs have better ears and noses; hawks have better eyes; bears, wolverines, and big cats have superior claws; horses and antelopes are faster; and so on. Even our teeth are adapted to an omnivore diet—front teeth for biting fruit and meat, rear molars for grinding nuts and plant fibers.

Rather than tailor ourselves, or wait for evolution to do it, we invent objects to help us overcome these shortcomings. This runs from atlatls, the throwing sticks that helped hunter-gatherers throw spears harder and faster, to binoculars to see the far horizon. When accident or disease results in a disabling physical condition, it's just another shortcoming.

Peg legs, wooden arms with hooks, glass eyes, false teeth, hearing horns, and eyeglasses are traditional examples of what we now call synthetic replacements, or "prostheses," for damaged or missing parts and functions of the human body. Medical prosthetics is a field with a venerable history. George Washington wore false teeth carved of ivory or wood, and Ben Franklin wore spectacles.

The early devices were all clearly false, and temporary.

But as they were improved in the last few decades, they became smaller, more intimate (contact lenses and some hearing aids), more permanent (tooth implants), and harder to detect.

The range of available implants is expanding constantly, blurring the line between technology and biology: stainless-steel hips, knees, and shoulders, random bits of metal (titanium rods in lower leg bones, metal plates on skulls), heart pacemakers, insulin pumps, colostomy bags, drains in ears, heart valves, steel mesh plates to replace bone, implantable teeth and lenses.

With a rapidly expanding senior population in the developed world, the number of cataract operations and lens implantations has skyrocketed. Over two million devices are implanted every year, and by 1988, more than eleven million people in the United States alone had permanent medical implants.

Arm and leg prostheses look increasingly like flesh-and-blood limbs. As far as social interactions go, the more invisible the implant or attachment, the better.

Most people are not comfortable around individuals in wheelchairs, those wearing artificial devices, or blind people with dogs or canes. Old impulses, buried deep in the human psyche, trigger social awkwardness and avoidance around those who are "different"; in this case, people with visible prostheses. At social gatherings, we pretend they're invisible, or we pepper them with questions about their arm or leg or whatever, then move quickly on to others like ourselves. To the wearer, it must seem that the devices are wearing them.

Eyeglasses are the notable exception, being entirely too common to notice, and only children are amazed by sets of false teeth. It wasn't so long ago that people with hearing aids were routinely left out of conversations, but the tiny new electronic models are worn entirely in the ear, and are largely undetectable. Keeping a person with disabilities socially connected is what the new electronic implantables and prostheses are all about.

Today, multiple amputees don't have to while away their lives in wheelchairs in medical institutions. Increasingly, they are entering the work force by telecommuting from their homes, or even holding down jobs away from home. To paraphrase an old John Wayne movie, when you call these people "handicapped," you'd better be smiling.

Getting a Leg Up: Limb Prostheses

Smart Legs

In 1998, Tom Whittaker did what few people have done: successfully reached the top of Mount Everest. And, oh yes, he has only one foot of flesh and bone. Researcher Hugh Herr would like to make it easier for other amputees, like himself, to climb hills and trudge through snow. He is working on a "smart knee" at Massachusetts Institute of Technology's Leg Laboratory.

Today's partial or full leg prostheses are pretty dumb mechanical devices, unable to respond to changing conditions, such as moving from hard flooring to carpet. Natural legs quickly sense the difference in resistance, and muscles adjust walking speed or force. A truly "smart" artificial leg would be able to do the same.

Today's prostheses boast battery-powered knee joints to allow for easy movement, something like power steering. One company even offers a leg with a hydraulic knee controlled by microprocessor. Linked to sensors that continually measure the position of knee and leg, the chip allows the artificial leg to mimic the motion and gait of the rest of the body. This technology helps the wearer change gait speeds, move from a smooth, hard surface to a rug, and climb slopes or stairs.

Advanced artificial limbs feature comfortable, cushioned sockets, molded closely to fit the remaining leg remnant. Most use "smart plastics" in the socket that adapt to and remember the shape of the residual limb they have to fit. This is intended to eliminate pressure points, and the close fit enables the wearer to better control the movement of the leg. Still, to achieve balance and natural walking or running, the brain needs the feedback from the missing muscles and bones.

But soon, wearers of leg prostheses may be able to feel the ground beneath their artificial feet. At least one manufacturer, the Hanger Orthopedic Group, Inc., is developing pressure sensors for leg prostheses. The Sense of Feel Sensory System is an effort to restore that feedback from the ground to amputees.

Pressure sensors in the sole of the artificial foot send tingling signals to the amputee's residual limb. The more pressure exerted on the artificial foot, the greater the sensation in the residual limb. The amputee's brain soon interprets the sensations in the residual limb as being from the foot, not from the stump, a phenomenon known as cerebral projection. It's definitely a step along the way toward really smart prostheses.

Smart Arms

The Motion Control Utah Arm and Hand is a removable "smart" prosthetic available from Motion Control, Inc., in Salt Lake City. It's a far cry from Captain Hook's wood-plus-hook replacement hand. The result of ongoing research funded by the National Institutes of Health, it is a myoelectric prosthesis, a device run by muscles plus electricity.

The unit is made of a sophisticated nylon composite material, strengthened with carbon fiber and fiberglass that is about the same weight as an arm of flesh and bone. It also looks a lot like a "real" arm. Custom-fitted to the remaining upper arm, the device runs on a self-contained battery that replaces muscle power and enables the elbow to move easily.

Electrodes in the socket are close enough to the wearer's upper arm muscles to detect faint electrical impulses in those muscles. Amplified by the electrodes, the signals are then transmitted to the prosthesis and used to guide the mechanical hand.

The Utah Arm's function is not quite so seamless as the fictional one Luke Skywalker receives at the end of *The Empire Strikes Back*. It also cannot punch through walls, like the Terminator's. Wearers control the hand by consciously flexing and releasing their upper arm muscles, something that takes considerable practice.

After a period of training, however, they can again do many two-handed tasks around the house and even in outdoor jobs.

Artificial hands may look fleshlike, but the skin does not feel anything, so there is no feedback. That means the wearer has to keep looking at the hand to see what it is doing. This is very different from the normal situation, because fingers, especially the tips, are densely supplied with nerve endings for touch and temperature, and a large area on the surface of the cerebral cortex is involved with receiving and interpreting the information.

Under development by Hanger Prosthetics and Orthotics are prosthetic hands with microsensors in the fingertips that respond to hot and cold, and pressure sensors. Both types of sensors send signals to electrodes on the wearer's skin where the prosthetic is attached. The pressure sensors produce a "tingle" response on the wearer's skin. More pressure increases the "tingle" response, so the wearer can tell how tightly or loosely an object is being held.

The temperature electrodes grow warm or cool depending on the signal from the fingertip microsensors, restoring to the wearer the ability to sense temperature with their hands.

New Expectations in Amputee Sports: Toward the Bionic Man

You've seen the pictures in the news: guys in wheelchairs playing basketball. Do you remember the woman leading the hike up the canyon—the one with an artificial leg? And the next time you're on the ski slope, look around and notice the fellow with one leg and tiny skis on the end of his poles.

Increasingly, people who have lost arms or legs are picking up all facets of their pre-accident lives using the latest generation of prostheses.

Amputees are pushing the envelope of the possible, in work and recreation. Today they have a range of leg styles to choose from, and some use more than one model, depending on the activity. Golf for a leg amputee seems reasonable, but what about skiing, rock climbing, running, dancing, biking, or sailing?

Almost ignored until recently, interest in amputee athletics has grown tremendously over the last ten years. In the 1970s and 1980s they were the butt of countless jokes, but now the culture seems to get their point: instilling a sense of self-worth among those who have taken a major reversal.

Following the 1996 Paralympic Games held in Atlanta and the 2000 Games in Sydney, a growing number of amputee athletes have started training to compete in athletic events themselves. Some of these people were athletes before their accidents; some were just young and strong and determined not to let one unfortunate accident slow them down.

The Paralympic Games are the world's second largest sporting event, second to the Olympic Games. More than four thousand athletes from approximately 120 countries competed in the two-week-long Sydney event in October 2000. The program included the normal running, jumping, throwing, and swimming events. Participants are world-class athletes, although so far their times and distances are well off the world mark for their sports.

But that could change.

Could a double amputee wearing very smart, high-tech legs with superior recoil someday be faster than nonamputees? Why not design arm prostheses to be more efficient in the water than arms? There are already "racing models" of artificial legs—why not arms to propel swimmers to faster finishes?

Today's paralympic athletes, demanding high-performance artificial arms and legs, are taking the first steps toward the Six Million Dollar Man and bionic woman of tomorrow.

Electronic Nerves

Ever since Puss in Boots' seven-league boots, people have dreamed of being able to run faster than humanly possible. The first modern superhero, Superman, can "leap tall buildings at a single bound," besides being able to fly by no visible mechanical mechanism at all. Okay, it's a metaphor. Taking it semirealistically, though, all this is clearly impossible if we are confined to flesh and blood, but what about using bionic assist?

The bionic man and woman from the old TV series had super-speed, but we never saw the nuts and bolts of how it was accomplished. The zany clay animation film from Nick Park of a few years ago, *The Wrong Trousers,* postulated a pair of unstoppable robotic trousers.

But seriously, how close are we to superspeed, or even walking robots? Human walking, it turns out, is very difficult for a mechanical device to master. Walking on two legs demands movable joints, a pelvis, precise coordination among major muscle groups in the legs, and the action of stretchy tendons. If any of these components is missing or diminished, people have various problems with mobility, or can't walk at all.

To achieve upright balance, the body has gravity sensors in the inner ear (the cochlea) and mechanoreceptors (stretch receptors) in skeletal muscles. Together they tell the brain which way is up, and which muscles are working, and enable it to program the legs to walk. At a minimum, eight leg muscles are required to stand; another eight are needed to walk. Graceful walking requires the help of even more muscles. Walking is actually a series of short forward falls, catching the body just in time.

The balancing required in upright walking is still difficult for the human brain, even though our ancestors started doing it several millions of years ago. It's easy to lose the ability to balance if the muscles aren't exercised regularly, and falling is one of the most common symptoms of old age or failing health.

David Reinkensmeyer at University of California, Irvine, is developing a robotic harness that helps the nervous system recover arm and leg movement ability in those with neurologic injuries, such as stroke and spinal cord injury. An intricate set of "mechatronic" devices guides them to relearn walking on a treadmill, giving them only enough strength to move. Such "rehabilitators" help us to understand the adaptive control processes that enable motor learning. We seldom think of robots as a carapace around us, but

Reinkensmeyer's are. A single walker-helper is like teams of people who (usually rather clumsily) assist the injured. The great talent of a helping harness is that it is both smart and intimate.

Professor Reinkensmeyer's laboratory develops robotic devices for manipulating and measuring movement in humans and rodents. Instead of costly nurse teams, usually four nurses to one patient, these devices give mechanical assistance in retraining arm movement following stroke. They can provide movement training remotely over the Internet. Such help promises to be among the first widespread medical 'bots.

It's extraordinarily difficult to construct a mechanical device that can walk smoothly on two legs. In Japan, the Honda Motor Company spent more than a decade, and millions of dollars, on a robot that can walk as well as climb and descend stairs. But that's all it can do, and other robotics experts wonder if it was worth all the trouble.

Joe Engelberger, one of the founding fathers of robotic devices, feels that wheeled robots are the most practical, given the design problems of walking ones. Many other roboticists agree with him.

Nonetheless, the work at MIT's Leg Lab and elsewhere continues, in part because of the interest in helping amputees and paralyzed victims of spinal cord injuries. Designing mechanical devices that walk naturally helps the scientists understand how a person walks. Then they can help create better prostheses for amputees, as well as sensor and control systems for paraplegics. As well, humans aren't the whole point. Many smaller robots will need to walk across impossibly rough terrain or even, like geckos, up walls. These uses will transcend humans by doing tasks we cannot, justifying legs as a "new" capability.

Replacing Nerves with Neural Prostheses

People with nervous system diseases or injured spinal cords have different challenges than do amputees. They have all their body parts, but cannot move (or often even feel) them because the connection between brain and muscles has been disrupted or destroyed.

The spinal cord is a bundle of nerve fibers running inside the vertebral bones of the back. The signals they carry connect the brain with the rest of the body. Sensations from skin and muscles travel up into the brain (sensory nerve input to the brain), while signals controlling the movement of muscles travel downward (brain output to motor nerves).

In a person with a damaged spinal cord, the brain no longer receives signals from the mechanoreceptors in the muscles, and it can't issue any commands back to the nerves that control the muscles. For people with damaged spinal cords, the extent of their disability depends on how high up the spinal cord was damaged. Generally, any part of the body below a break or a lesion will be affected. Paraplegia results when the damage is in the midback region, below the shoulders. These people can move their arms, but their legs are paralyzed.

People with broken necks are quadriplegics, unable to move either arms or legs. A common accident occurs when a person dives into too-shallow water, hitting a pool bottom or sand bar underwater, breaking his or her neck. Those unfortunate accidents usually leave the person wholly or partially paralyzed from the neck down. The late actor Christopher Reeve was quadriplegic after breaking his neck in a fall from a horse.

For both paraplegics and quadriplegics, a pair of related technologies, functional electrical stimulation (FES) and brain computer interface (BCI), are being developed.

Functional Electrical Stimulation

The most developed system is FES, for "functional electrical stimulation." It consists of implanting electrodes under the skin next to the unresponsive muscles, in effect taking the place of the patient's damaged nerves. The first commercially available FES, or neural, prosthesis to restore movement to a paralyzed arm is called Freehand, by NeuroControl Corporation of Cleveland, Ohio.

Using the Freehand, the wearer's paralyzed arm can again pick up a fork, grasp a soft-drink can, or write with a pen. In order to use the device, the patient must be able to move his or her shoulder. The

Federal Drug Administration estimates that more than fifty thousand quadriplegics in the United States might be able to benefit from this electronic implant.

Moving a small "joystick" attached to the shoulder sends an electrical signal to a wire coil strapped onto the person's chest. Just beneath the wire coil is an implanted computer chip that acts like a radio receiver to pick up the signal. The chip relays the signal through wires under the skin to small metal plates called electrodes. The electrodes are implanted in the arm muscles (usually on the opposite side of the body from the joystick), and they conduct the signal to those muscles. The idea of the whole relay system is to replace the damaged nerve pathway.

As with the Utah Arm prosthesis, the wearer must learn to convert upper arm or shoulder muscle movement into fine hand movements. Training allows the wearer to perform many normal hand and arm movements, like throwing a ball, eating, or writing with a pen.

For legs, there are similar systems. Centers in Cleveland (Department of Veterans Affairs Center of Excellence in FES and Case Western Reserve University), Miami (University of Miami Medical Center's Project to Cure Paralysis), and Augusta (Neural Engineering Clinic and MaineGeneral Medical Center), among others, have been working toward the goal of restoring the capability to stand and walk since the 1960s.

The first FES systems were not implants. External electrodes in pads were placed over leg muscles, causing them to contract when stimulated electrically. This allowed the wearer to exercise, avoiding the typical wasting of paralyzed muscles. Hundreds of patients have used these devices, and among them some can stand, and others can walk.

About twenty years ago, researchers started implanting wires under the skin, with the electrodes ending directly in one or more of the leg muscles. These systems were also successful, and allowed for more precise control.

More recent models use a computer chip with multiple electrodes. One such model, the Nucleus FES22, is an implanted chip with twenty-two wires that directly stimulate a series of muscles that control standing and walking—from the foot through the buttocks. The system is powered from outside by induction, similar to an electric toothbrush. Power is transferred to the implanted chip through a circular antenna positioned directly over it on the skin. The antenna, in turn, is connected by a wire to a battery-powered control box small enough to be worn on a belt around the waist. The wearer controls his movements by flipping one of several switches on the box. For different movements, a different sequence of muscles is activated.

Over the years since the first implant, the component devices have gotten smaller and more responsive. One refinement involves the sensors, tiny gyroscopes to sense balance and direction, and accelerometers that sense movement. Originally used for air bags in cars, they are similar to sensors in advanced prosthetic legs, and replace information that used to travel to the brain from stretch receptors in the muscles. The sensors allow the system to detect if the knee is straight or bent, moving or not. That information is used by the control box computer to stimulate the leg muscles so the person can continue standing.

What is it like to wear a FES device? Jennifer Penko was a young adult snowboarder in New Hampshire when she ran into a patch of ice that changed her life.

INTERVIEW WITH JENN PENKO AND TIM FRENCH, HER FIANCÉ

(Penko) We were snowboarding and I had hit a patch of ice and went down a thirty- or forty-foot embankment, kissed a bunch of trees, and ended up facedown in the snow.

Q : When was it clear that it was a spinal cord injury?

A : (French) When the rescue professionals started asking if she could feel this or move this, and she couldn't move her legs and stuff, it was tough. In the smaller hospital in upstate Vermont waiting for the helicopter to bring us to a larger hospital, it was then made clear that it was a spinal cord injury and it was very serious.

Q : What made you think that there would be something to the idea of muscle stimulation?

A : (French) Muscle stimulation means being able to stand, possibly being able to walk with the more advanced system that could be in the works further down the line, instead of just sitting back and doing nothing about her injury. We just pursued whatever was available to us and the most opportunity was right here in Cleveland.

I was most happy for Jennifer when I first saw her stand. To see the smile on her face, and to see how thrilled she was to be standing is what I think really made me happy.

Q : When you turn the dial [on the FES], what do you feel?

A : (Penko) I have an "incomplete" spinal cord injury, which means that I have capabilities below the injury point. One of my capabilities is sensation, so when I hit the button and three seconds later the electrodes go on, I can feel them going on, and I can feel my muscles contracting when I stand up.

Q : How is it different from an able-bodied person?

A : My muscles are fully contracted. It is a hard, hard contraction; therefore, the muscles will fatigue quicker than in an able-bodied person. I also still need my arms for balance and to help lift me up out of the wheelchair.

Q : Has the implant changed your life?

A : Sure. Before the surgery I had to use long-legged braces to stand up out of the chair, and it would take me fifteen minutes to get into them—with help. And then I would be able to stand at most an hour or two in an entire day.

With the FES system I gained not only the medical benefits of standing out of a wheelchair, but I also get the functional use out of it. I can stand by myself, and reach and stand one-handed. I can ambulate around a lot easier than I used to, so it has made me more independent.

I am home alone a lot and I couldn't do that, be independent, reach things and do things without the implant. I would have to rely on a caretaker to come over and help me.

Q : So weigh the cost and benefits. Was it worth it?

A: The cost, of course, is the surgery and the therapy and the cost of the FES system. The benefits are independence and prevention of a lot of medical conditions that happen when you are in a wheelchair a lot.

Q: How does it feel to stand up with this system?

A: The first time I stood up was the most liberating experience of my life. For anybody in a wheelchair, it is just incredible to be able to stand again. It is fantastic.

Q: How far do you hope to go with it?

A: I hope to get to the point that I don't need to rely on the walker as much as I do. I would like to be as mobile as possible. I would love to be able to just pick up the system and go anywhere, and I can now, but there is a little bit of limitation with the walker and stuff.

Q: What about walking?

A: They are still working on that, but I am walking with a walker, if you will. I kind of do a two-step or a swing-through with my legs. I don't bend my legs or anything, but I do walk.

Q: Is your experience with the FES device unusual?

A: This is a research study, and everybody is different. There are people in the study that have "complete" injuries and they are able to stand. You really don't know what your sensation, what muscle groups you are going to gain or what you have and what you don't have.

   When that sensation that comes back up to you—*wow!* Looking down, I can't control those muscles, but still, it is a lot more than what I could do before. Before, I would look down and see flaccid muscles within a brace. Now it's my *own* muscle holding me up—one step closer to standing like a normal person.

Q: Do you feel different, like a cyborg?

A: The implants are like any others. I mean, if someone has a pacemaker do they feel like, "Oh, I have this pacemaker, and I constantly feel like it is something that is protruding and alien-like?" Because this whole system is underneath the skin, you in essence forget about it. You know that it is there, but it just becomes a part of you.

I have the implant in the hip and it sticks out a little bit so sometimes I think, oh it is alien . . . but it is not protruding and it is not invasive. So it really becomes a piece of you. Except for the control box, of course. But that's what allows me to do things. I need my wheelchair to get around. I need this box to stand. It's just another piece of adaptive equipment that I need to use to gain function.

Q : Are you working toward any goals?

A : My short-term goal is to walk down the aisle at my wedding.

Q : Do you think you'll be able to do it with the current system?

A : I practice every day. I want to walk up the aisle and stand for the entire ceremony. And walk back down again.

Q : How far are you from that goal?

A : I think I am further along in walking down the aisle than I am for the plans for the wedding, so I am pretty confident that I will be ready once the time comes.

What about the future? No matter how miraculous they are to those in need of their assistance, today's devices are crude compared with what they might be in the future. From everything that we know about technology, we can expect that future FES systems will be smaller and more powerful than the current equipment.

Wireless Technology

The idea of wires running through the skin triggers the "yuck" reflex in many people, potentially limiting the usefulness of electronic aids for people with less severe injuries. Stroke victims with just one arm paralyzed, for example, might resist such an invasive system.

Just over the horizon, however, is wireless technology that would enable tiny implants to radio data to an external computer. The devices themselves would be powered by magnetically inducing an electrical current from outside the body, like an electric toothbrush. The external antenna and power source would be worn over the implants. Philip R. Troyk of the Illinois Institute of Technology has already constructed wireless sensors no larger than a Rice Krispie.

Conceivably, sensors the size of rice grains or smaller could be implanted in a patient's muscle through a large-gauge syringe in the doctor's office. The devices could pick up local nerve signals and relay the information via radio to an external computer. They could also receive power and send out the mild shocks that stimulate the muscle into action.

The next step would be to shrink the control and power box to the dimensions of a palm-size electronic device, or embed the controls in clothes, like some wearable computers.

Where could this go? For normal people, electronic augmentation may allow better coordination of existing muscle groups. This frontier is just opening, but it promises to overcome, for example, muscle fatigue and disorientation in tough surroundings. A fireman who needs to momentarily increase his strength could use such an electrical pulse to avert danger or save a life.

CHAPTER 2

The Augmented Animal

Can technology save us from our faltering natural machinery?
Smell better, see better, overhear conversations—
and throw away those specs.

*For those societies that already provide health care services to their populations, it's not
a big step . . . to provide some particular kinds of enhancements as well. That is a somewhat nervy
prospect. . . . The difference between that and previous government interventions in genetics
could be that it's not coercive—that it's an offer of a subsidy.*

—Peter Singer

Improved People

You're playing your normal game of tennis, when a couple you've never seen before come onto the next court. They're about the same age as you, but their first warm-up strokes tell you that they're in another class entirely. Volleys rocket across the net, barely nicking the back line, then return with equal ferocity. The couple move with uncanny agility, picking up lobs and net shots that look nearly impossible. But they're not buff, or younger, and don't seem much different from you.

Then you get it: they've got those new neuromuscular add-ons. You read about them in some newspaper article—pricey, but apparently they work. You watch their shots zipping back and forth for a moment and your own partner comes over. "Tired?" You realize that no, the sensation you feel is not fatigue, but a vague sort of discontent, as if you had already lost the match. And in a way you cannot define, you have. Game's over.

This hasn't happened yet, and may never come to pass, but it seems plausible. The technology to stimulate muscle growth lies at least a decade or two away, but the drive to make it happen has been with us in a general way for millennia.

Does anyone in your family have chronic back pain? Has a relative gotten a pacemaker to regulate the heart? Then your family is already on the threshold of the bionic future. Clearly there is room for improvement in the human body. Augmentations can help shore up parts that grow frail and weak with age.

Consider yourself, starting with your head. It is conspicuously unprotected, and you can be killed by a simple fall. What if the body stops breathing? Starve the brain of oxygen, and within minutes, it can be damaged forever. An added system, perhaps riding like a backpack, could go into pumping more oxygen from a fixed store into the brain's circulation, stimulating it. For the elderly this would be a crucial aid, especially since one of the prevailing afflictions of old age is a diminished oxygen level in parts of the brain. This deficit brings on loss of memory and function.

Just below the brain, our spines are also vulnerable to aches, pains, and injuries. Spines are interlinked bony columns riding on discs of cartilage. Plastic movement comes from this array, but with age the discs flatten under impact from walking or running, allowing the stacked vertebral bones to butt against nearby nerves, sending piercing pain to the brain above. This is a source of distress for a large fraction of the population. Strengthening this system from outside seems difficult, although we are already replacing some body parts elsewhere with fairly strong artificial materials, mainly restoring joints like hips and knees.

Our vulnerability springs from natural selection. Our skin

repairs itself, but our kidneys do not. Skin is apt to be bruised, torn and cut, so nature found it worthwhile to evolve repairing mechanisms. Deep organs are less likely to be hurt, and if they are, the damage will probably be fatal quite quickly, since the maintenance mechanisms of the body must run steadily. So to replace inner damage, we must use artificial organs, because nature thriftily preferred simply to let individuals die rather than develop the means to fix large, deep organs.

This opens the door to artificial organs, whether they ride within the body or without. So far artificial hearts, kidneys, and livers are rather bulky machines that function well only outside the body, immobilizing the patient. That may change within a decade, most probably with the vital heart. Transplants have so far made the need for such technology minimal, but the shortage of donors, plus rejection problems in the body's immune systems, seem to set a ceiling on this biological technology.

Beyond the heart, the other important organs seem simpler. Kidneys and liver are filters. The pancreas and thyroid secrete chemicals for overall body regulation. These functions seem approachable in devices. Such machine organs seem to be the next major step for cyborgs or androids or humans alike.

Where might this lead? Where are we today?

Better Innards

Heart Surgery with a Screwdriver?

> I could stay young and chipper
> And I'd lock it with a zipper,
> If I only had a heart.
> —The Tin Man, *The Wizard of Oz*

A routine checkup for a heart patient in the future might well be conducted with screwdriver and battery tester instead of with a stethoscope.

Attempts to build artificial hearts started in the 1960s, resulting in two well-publicized trials of the Jarvik-7 device. Unfortunately, the

results were so unsatisfactory in terms of quality of life for the patients that the idea was all but abandoned. Today, the Jarvik-7 is
used as a temporary left-ventricle-assist device for patients awaiting heart transplants, and a newer, much smaller, implantable device is in human trials, along with several other devices from
different research groups.

The left ventricle is the muscular chamber that pumps oxy-
genated blood to the brain and the rest of the body, doing about 80
percent of the heart's work. Patients with heart failure suffer from a
thinning of the heart muscle and reduced pumping capacity. With
the existing external devices, patients can be kept alive for months
or even a few years as they wait for a compatible heart to become
available from a donor.

That waiting time can be considerable: only two thousand hearts
a year are available in the United States for a growing pool now
numbering about a million patients with end-stage heart failure. The
aging of the population will only increase the demand. And even
when a heart transplant is performed, the patient must take potent
immune-suppressing drugs for the rest of their lives, increasing risks
from infections and cancer.

Artificial hearts could solve both those problems. So how long
will it be before doctors have the option of implanting an entire mechanical heart, or even just a small assist pump?

Several devices are on the horizon; some are in animal trials,
and a few have already been implanted in human volunteers.
Although they differ in design and construction, they share similar challenges: They must be small enough to fit within the chest,
materials must be nontoxic and biocompatible, the pumping action must not damage blood cells or cause them to form clots, and
they must work for many years without lubrication or maintenance. The new heart should also be able to respond to demand
by pumping at different speeds. Then there's the problem of an energy source.

We are so used to having a pulse that we expect full artificial
hearts to mimic the pumping action of the natural heart. But continuous rotary pumps are also under development.

Nature solved the problem of moving fluids around the body with a type of muscle cell that twitches spontaneously. Connected into clumps or sheets, all the cells twitch at the same time. In embryos, the heart develops as a tube, like a thick-walled blood vessel. This tube soon divides into two tubes side by side, then twists itself into the fist-shape of the adult heart. To pump blood, the entire organ alternately squeezes and relaxes, and a series of one-way valves allows the blood to pass through the four inner chambers.

But an artificial heart pump might not have to work in the same way. Biological organs are constructed of living tissue, so to nourish and repair the cells in that tissue, all parts of every structure must be connected to the rest of the body so they can be reached by blood vessels. That's a biological engineering problem, and the reason why there are no free-spinning organs in nature. There's no way to get blood to the tissues.

A mechanical device has no such constraints.

In the future, artificial heart recipients may have quiet, continuous-flow rotary pumps in their chests. Some of these may be simply left-ventricle-assist devices, permanently installed within the ventricle. Some may be full heart replacements. In either case, scientists will need to investigate whether blood vessels in the rest of the body need the rhythmic pulsing of blood flow to stay healthy.

As to energy sources, long-life batteries in some devices can be rechargeable with a wireless energy transfer through the skin (like cordless electric toothbrushes). In others, the batteries are external, connected to the device via implanted wires.

Heart pacemakers already enable tens of thousands of people to lead normal lives with hearts that cannot beat properly, or at all, on their own. Within a decade, we can look forward to most of the technical hurdles of artificial hearts having been settled with the same amount of success.

Other Artificial Organs

Although we may not usually think of it, the skin is the body's largest organ. People who are burned over much of their bodies experience catastrophic fluid loss over the short term, and an extended, very difficult, and painful recovery. Often they are extensively scarred even after a "successful" recovery. Research into artificial skin has been under way for decades, with limited success.

Recently, however, at least two different types of temporary, synthetic skin have been approved for human use. One uses a nylon mesh base into which cells are grown from the foreskins of circumcised babies. It is applied like a sheet over burned skin—the human cells contain proteins and growth hormones that help the burn victim's new tissue grow.

Another type uses a meshwork of animal collagen, a substance that also helps human tissue hold its shape. This mesh is applied to the burned skin, covered with a layer of silicon to keep fluids in. Once in place it encourages any remaining skin cells to grow into it and form new skin.

So far, the replacement skin looks a bit odd, mottled, and at times shiny. But improvements are coming rapidly.

Both types of artificial skin, as useful and revolutionary as they are, ultimately rely on the burn victim's own skin cells to regenerate and replace the lost skin. Eventually the temporary synthetic skin is broken down by the new tissues.

Skin is crucial to us because it is literally the face we present to the world. Artificial organs inside the body have little immediate social impact and can be secret. This suggests that athough skin substitutes are perhaps the most developed, interior organs will overtake them in use.

All this will play out in the next decade, as progress quickens. For the moment, a transplant from another human or cadaver is still the best answer for a failed organ. Functioning, permanent artificial organs await another level of medical understanding and research.

Eavesdropping on the Future

Suppose you have never heard a dog bark, the wind sigh, or the ocean's stormy roar. You can see these things happening, or infer them, but the world of sound is like colors unseen. The serenity of such a life might be valued, even sought after— but not by most, who would rather listen to Mozart, or a baby's cry. Yet it is in the declining years of life, ripe with memory, when our hearing begins to fade. Worse, it distorts, and we lose our highs and lows. Yet we remember.

The cochlea is a tiny snail-shaped organ in the inner ear containing sensory cells called "hair cells." Minute hairlike projections on these cells detect sound waves and relay the information to adjacent cells of the auditory nerve. The nerve cells convert the information to electrical impulses, which travel to the hearing center of the brain. Damage to the hair cells impairs hearing.

Under development since the 1950s, cochlear implants have recently become widely available. In 1984 such an implant became the first artificial implanted human sense. A device is worn on the ear, similar to a hearing aid, called an external processor. This external coil of wire detects and delivers the noise signal to the internal coil, which is implanted under the skin. From the internal coil, wires run to the inner ear, ending in electrodes that stimulate the cochlear nerve cells directly. This setup bypasses the damaged or absent sensory cells, reproducing recognizable speech.

The external processor also delivers the electrical power for the internal electrodes. The internal devices work on the physical principle of induction, when a changing outside electrical current induces weaker currents in the implanted wire coil, to deliver a signal. The external power/signal coil is kept centered over the internal coil by including a magnet in the internal coil that serves to anchor a matching magnet in the external coil.

Hearing persons who have recently become deaf can learn to understand more than 90 percent of spoken words with any one of several electronic ears on the market. Other patients have somewhat less success, but all can hear better with the implanted electrodes than without.

As promising as these devices are, some researchers are looking for another way altogether to improve hearing.

Hearing is diminished when the tiny "hairs" suffer damage from loud noises or diseases. Injury done at rock concerts of the 1960s only now manifests itself widely in the baby boomer generation. Researchers are looking for ways to regrow the ear's hair cells. Other animals, including sharks, zebra fish, and chickens, can produce new hair cells after an injury, but not humans. However, some research indicates that certain growth factors (chemicals) may cause nerve cells in the cochlea to multiply. A new kind of implant might continuously release these growth factors into the ear, inducing nerve cells to repair themselves or produce new, healthy cells.

Some researchers are even thinking of bypassing the ear altogether, implanting arrays of electrodes directly on the surface of the brain. Those new bionic ears await the development of tiny computers and materials that the body will tolerate. But no one is willing to say it's impossible anymore.

Rights and Resistance

The development of the current crop of cochlear implants offers a cautionary tale for would-be developers of new medical prostheses. For years, doctors and researchers in the field of hearing impairment tenaciously resisted the technology from its inception. They questioned the theory behind the devices, maintaining that it would distort sounds and disorient the users. All of that contrariness was theoretical, objections made without examining actual clinical results. When positive results became clear, this opposition waned. But the initial response is instructive. We have a natural resistance to change, particularly change done to our bodies. But the story of hearing technology does not end with clinical trials.

Ironically, now that the implants have been shown to work, representatives of what is called the "culture of the Deaf" in the United States continue to resist their use in children. They contend that deafness is not a handicap, but part of human diversity, and that it should not be "cured" in children who are too young to make the decision on their own. Unfortunately, this ideology conflicts with years of research results establishing that language is best learned while young.

Throughout the development of the devices, however, there has been no shortage of patients willing to try them. The Deaf culture advocates represent only a small minority of deaf people.

It is tempting to speculate whether there is something peculiar about deafness or hearing impairment that causes these reactions. Whereas few people resist glasses or contacts if their eyesight is not good, many people refuse to admit that they can't hear well.

Some assert that not being fully tuned to the world's cacophony is a blessing. The science fiction author John Varley depicted an extreme state of this attitude in his memorable 1978 novella, "The Persistence of Vision." The people in the story find the double truncation of both sight and sound not isolating, but rather, spiritually engulfing. Varley conveys an experience that, once fully felt, has a hypnotic quality. This event can lead a sighted and hearing person back into the heightened involvement that deprivation forces upon the mind. It is an eerie and oddly memorable idea, though not truly plausible.

In terms of new medical treatments, often what is seen as experimental and outrageous, even unethical ("Man should not go there!") quickly transforms in the face of success. Many come to see that they have a responsibility to act for the next generation, if the technology works. In a complete turnaround, what was regarded with hostility becomes the norm. Parents face criticism if they do *not* opt for the technology for their children.

Vaccines against childhood diseases were regarded with mistrust when first introduced, but once their safety and benefits became established, society *expected* parents to do what was right and have their children vaccinated. In fact, vaccination is a prerequisite to enrollment in U.S. public schools.

The few parents who still resist must have established religious reasons, and are considered well outside the mainstream of society. Unvaccinated children still die of measles.

Other medical treatments have similar histories. At first they are considered controversial, and medical insurance will not cover the procedures or drugs. It is worth remembering that anesthesia was denounced as harmful because pain was once thought (in the 1840s) to have healing effects. This opposition faded quickly, because people have a powerful aversion to pain.

Issues like protection from disease are rather more distant and theoretical. But as the new methods subsequently become the standard treatment, it is expected that everyone will avail themselves of the treatment.

In extreme cases, parents are forced by the courts to allow their children to be treated, even against their desires. For example, in September 2000, the English government forced a couple to allow doctors to surgically separate their conjoined twins, with the foreknowledge that the weaker twin would die on the table. The alternative was for both twins to die a few months later. The parents' religious beliefs did not allow for medical intervention in such a decision, but the state intervened, arguing from a social utility view that at least one would live.

In such painful cases, society usually works with a rough-and-ready calculus of utility. Better one live twin than two dead ones, public opinion holds, and the legal system follows suit. Where there is no good solution for everyone, most people simply go for a maximum benefit.

There are implications here for any technology that insists that bad effects be weighed against good. Most technologies demand something of us, if only the discomfort of putting them into the body, and the time to adjust. Usually judgments are left to those who would gain from them, but for children or the impaired, we must go through a balancing act. This aspect of our technological landscape will not change, no matter how wondrous future devices may be. Cyborgs will face a trade-off between the risks of altering the body, and the potentially large payoffs. Voices on both sides of the issue will call for state intervention. Only when the benefit becomes quite obvious will conventional wisdom come down decisively for intervention and implantation.

In the end, utility wins. Society insists.

Hearing Beyond Limits

What do dogs and elephants have in common? They can hear sounds most of us can't. Dogs' ears can detect much higher-pitched sounds than ours can, which is why the very high-frequency sounds emitted by dog whistles are undetectable to us. While a healthy young human can detect sounds between 20 and 20,000 cycles per second, or hertz (Hz), dogs hear "ultrasonic" sound waves at frequencies up to 40,000 Hz. There is, of course, some variation in what people can detect. Some feel a tightness around the temples or even hear a faint, very high-pitched sound when ultrasonic burglar alarms are activated. Others sense nothing.

At the other end of the frequency scale, elephants rumble to each other using low-frequency sounds that humans detect only as a throbbing in the air. They cannot be detected by the human ear drum because they are too large; the lowest notes sounded by pipe organs in cathedrals similarly vibrate the body itself, leading to sensations some find stirring, though unheard.

Within the inner ear's cochlea are a series of fibers like harp strings, known as basilar fibers, of different lengths. Only sound waves of the right size to vibrate one of these fibers are detectable. The fibers stimulate the specialized sensory "hair" and nerve cells that carry the stimulation to the brain, where it is deciphered as sound. The lowest elephant notes resonate with fibers larger than those in human ears. On the other end of the scale, deer warning "whistlers" mounted on cars and dog whistles emit sound wavelengths too short for the cochlear fibers to detect.

The electronic ear, however, could pick up sounds beyond the range of normal human hearing. Signals detected by the external processor are passed directly to the nerve cells, bypassing the filter of cochlear fibers. Adjusting the sensitivity of the device would make audible the world of unheard sounds. And what might those be? Since no human ear has heard them, the answer is speculative. But if dogs and deer have ears that hear very high-pitched sounds, the noises must exist in nature. Quite likely, the upper harmonics of birdsong, insects calling, or the squeaking of small furry animals lie in this range. Maybe the wind makes high-frequency sighs. Deep sounds might include the creaking of trees in a windstorm, rumblings of the earth itself during (or preceding?) earthquakes, waves pounding the seashore.

Nature has not selected for humans who can hear such warnings, because they made little difference when we evolved. But that does not mean they would be useless to us today. An augmented engineer might usefully hear a stressed bridge groan before it collapsed. Factory workers could hear high-pitched notes telling of machinery malfunctioning.

An entire symphony of unfamiliar sounds awaits its first explorers. We should be open to it.

Seeing Better

Imagine that you were born blind. Your world is then ruled by sound and smell and touch, and from birth your only way of knowing your mother's face is to feel it. Smell and sound tell you of things beyond your reach, but they are blunt sensory instruments. Animals can smell and hear more acutely, but in your world it is difficult to know even what this means, because by far our greatest fount of knowledge comes from sight. So wedded are we to our eyes that there is no clear, sharp division between seeing and thinking about seeing. Light falling onto our retinas is processed every step of its way to our brains. Editing and shaping of these impressions occurs all along the optic nerve. Sight gives us a gusher of knowledge. To navigate life without it is to risk a fall with every step, to risk life itself by merely losing the way on a path. Imagine a dark world without the perceptual reach of sight—to see to the horizon, to know in sharp detail what smell and sound bring only in crude, different ways. It is not without reason that the act of finally knowing something is termed, "to see the light."

Perhaps still worse is to have sight and then lose it. Yet many elderly face just that, as diseases like macular degeneration slowly rob them of the sharpness they once knew. From having trouble reading they descend to watching television, or just sitting, unable to make out much of

what the blobs and splashes of light around them mean.
Yet they remember, and regret.

According to a recent news article in the *New Scientist*, pop singer
Stevie Wonder, blind since shortly after birth, will have an experimental microchip implanted in his retina that might help him see
again. Led by Mark Humayan, a team of researchers at Johns Hopkins Hospital in Baltimore has implanted similar devices in over a
dozen blind patients, who have then been able to see spots of light
and simple patterns. Worldwide, at least half a dozen teams of researchers and doctors are working on electronic retinas and artificial eyes.

Vision in humans depends on light-sensing cells called rods
and cones (because of their shape) that are found in a thin layer of
tissue, the retina, lining the back of the inside of the eyeball. Light
entering the eye is focused on the retina by the lens. The rods and
cones react to the incoming light and convert it to electrical signals
that are transmitted to nerve cells making up the optic nerve. From
there, the signals travel to the part of the brain known as the optical
cortex, where they are decoded.

But about ten million Americans suffer from macular degeneration, a disease of aging, or retinitis pigmentosa, the leading cause
worldwide of inherited blindness. Both of these diseases cause the
deterioration of the retina through the loss of the light-sensitive rods
and cones. Despite having an active connection to the brain through
a functioning optic nerve, people with these disorders cannot see.
For them, an electronic retina offers hope for the future.

For over a decade, scientists at Harvard Medical School and
MIT have collaborated on the Retinal Implant Program, developing
an implantable electronic retina. The implant consists of a thin plastic film in which are embedded a hundred electrodes. Not light
sensitive in themselves, they electrically stimulate the healthy retinal nerve cells, bypassing the damaged rods and cones.

The blind subject wears a pair of glasses containing a small electronic camera that records the scene in front of them. This visual

information is converted to an electronic code by a signal-processing microchip. The chip conveys an image to the electronic retinal implant, the information riding on a laser beam, sent from a miniature laser also mounted on the glasses. The process is similar to the way a television picture is carried on a cable.

The retinal implant itself has two silicon microchips. The first is a tiny "solar battery," which receives the light from the laser. The second chip, the stimulator chip, decodes the picture information carried by the beam and transmits electric pulses to the nearby retinal nerve cells. These will be carried to the brain by the patient's optic nerve.

The solar battery chip also provides electric power to the implant. The implant is a thin, delicate film designed to rest on the inside surface of the retina, very near the optic nerve fibers that carry signals to the brain. Three subjects who have been implanted can see dots that correspond to stimulated electrodes. In these experiments the implant is temporary and is removed after a few hours. So far, a safe, long-term, biocompatible implant has not been developed, but several labs in the United States and Europe are hoping to succeed, with luck, within a decade.

The Johns Hopkins team has temporarily implanted over a dozen subjects, some with an array of twenty-five electrodes. With that number of electrodes, they could see numbers, letters, and basic shapes generated by a computer and transmitted to the electrodes in their retina. "Some of these patients haven't seen for four decades," says Humayan. "It takes a while for them to see the letters, but once they pick up even a single dot, they make quick progress from there."

The results are encouraging enough for Humayan to predict that within five years they will be able to restore basic vision in such people with a permanent device.

Direct Implants

The electronic retina will not be able to help patients suffering from diabetes or glaucoma, because these diseases damage nerve fibers in the retina that lead to the brain. For those patients, and for people blind from birth due to nonfunctioning optic nerves or malformed eyes, a direct implant into the optical cortex of the brain may be the long-term answer. This approach would bypass the eye and the entire visual pathway—in effect, substituting an electronic eye.

In one version of an electronic eye, the patient wears glasses containing light sensors (probably a tiny camera) that pass signals to a portable computer worn by the patient. Wire leads from this computer run under the scalp, and end in stimulating electrodes that penetrate the surface of the visual cortex of the brain. When the computer sends an electrical impulse to one of these electrodes, the visual cortex of the brain gets stimulated, and the result is the appearance of a bright spot in the black visual field of the blind eye. With enough electrodes, patterns could convey visual information to the user, the way individual dots in the newspaper merge to create pictures or words. At this stage, the researchers are not expecting natural eyesight, but simplified images, enough to help the subject avoid objects such as furniture in unfamiliar places.

These are crude beginnings, just as television was a vague, colorless image only eighty years ago. But often, advances in microelectronics can suddenly leap ahead. True artificial sight may lie only a generation away.

Beyond Human Vision

What do honeybees and heat-sensing "snooper" aids have in common? They can see what we can't. More properly, they can see *where* we can't—where in the electromagnetic spectrum, that is.

Humans see using what we naturally enough call visible light, the tiny part of the spectrum that our eyes can detect. The six million light-sensing cones in a human eye are excited by a range of different wavelengths of light. We perceive them as the rainbow colors—red-orange-yellow-green-blue-indigo-violet (ROY G. BIV)— and they are the only wavelengths we can see. On the violet end, the wavelengths are shorter and more energetic than on the red end. But just outside our visible perception are longer infrared rays (IRs), which we feel as heat, and on the other end of the rainbow, shorter ultraviolet rays (UVs), which cause sunburn. A bee's eyes are tuned to this end of the spectrum.

Flowers have 'evolved to take advantage of the honeybee's vision. Using special camera film and quartz lenses instead of glass (glass blocks UV wavelengths), scientists discovered that what we see as very plain flowers may have spectacular patterns visible only in UV light. Called flower guides, they act like landing patterns at airports for incoming bees and other insects, attracting them to the nectar within.

Heat-sensing optics sensitive to IR allow soldiers to track enemy movements at night by detecting white-hot exhaust from vehicles, or the yellow and red of a person's body heat. Even footprints glow dull red for a while after a person or animal has walked by. Using a special IR-sensitive film in an ordinary camera, energy company technicians can take a nighttime heat-picture of a house. By the brightness of the color, they can gauge how much heat is escaping from the doors, windows, and roof, and where insulation is needed.

Some luxury cars already include a night-vision (IR) device to alert drivers to warm-blooded hazards like deer in the road. Finally, sensors in satellites routinely take so-called false-color images of the ground with film and optics sensitive to infrared light. These images allow scientists to distinguish actively growing vegetation, like crops, from slower growing forest, information not available in photographs taken in visible light.

Unaided, human eyes cannot see flower guides or heat radiation, unless an object is so hot, it glows in ordinary light. But then we are day-active animals—what about animals that are active after dark? How many of them use IR-sensitive night vision?

Actually, none of them do. They all use other methods to enhance night vision. Cats and other nocturnal mammals, as well as deep-water fish, have a well-developed layer of reflecting crystals behind the retina called a tapetum. The familiar shine of a cat's eyes at night is due to reflected light from the tapetum. Hold a flashlight at eye level and go hunting in your garden at night. Myriad tiny lights flash back from the darkness. Many of these are spiders, but most night-active creatures' eyes have a tapetum—a layer of cells in the eye's wall that reflects light back onto the retina. This layer enhances vision in dim light, very useful for deep-sea animals and those who forage at night.

Other animals have more exotic ways to find prey in the dark, relying on hearing, not vision. A barn owl can pinpoint the faint rustling sound of a mouse hundreds of feet away, using a ring of feathers on its face to focus the sound toward its ears. Unlike satellite dishes, designed on the same principle, the owl can fine-tune the signal by moving the feathers, which changes the shape of its face. Also, its ears are asymmetrically placed on its head, so there is a slight difference in the time the two ears hear the sound. Bat echolocation is even more elaborate. The bat sends out a pulse of very high-pitched sound that reflects back to its large, sensitive ears from objects in front of it. The bat interprets these echoes for position and size information of objects in its path.

In fact, the only animals that have heat (IR) sensors are some snakes. Known as pit vipers, and including rattlesnakes, these animals have pits with sensitive heat receptors on the front of their face. Just as we can feel a warm source on our eyelids, the heat receptors of the pit vipers pick up IR radiation from warm-bodied animals. The snake does not "see" the animals, however, the way the night-vision sensors do.

As far as we know, in all of nature, *no* animals have full infrared vision. This absence suggests that there is a fundamental problem in IR detection that biology alone can't solve.

How *do* we see, anyway?

The color of light is determined by its wavelength. The color of paint is fixed by the pigments in it—chemicals whose atoms are arranged in ways that absorb one specific wavelength. For instance, red paint contains pigments that *absorb* other colors, but *reflect* red light. Incoming red light is absorbed by pigments in red-sensitive cones in the retina of our eyes, exciting them to send a message to the brain. Other cones are green- or blue-sensitive.

Although animals and plants contain an array of pigments that absorb many different colors, no biological pigments absorb wavelengths in the infrared, so animals don't "see" infrared. Why? Possibly because the pigments that can absorb those wavelengths are poisonous. For instance, the night-vision goggles employed by the military use thin sheets of indium arsenide—in other words, a chemical containing toxic arsenic. Possibly living systems can't easily evolve the cellular machinery necessary to handle arsenic without damage.

The take-home message is that there are limits to natural systems that we can't get around using biological tools. Still, using the artificial eye technology described above, humans could extend their range of visible perception. The tiny camera and implanted retinal chip could easily be made sensitive to other wavelengths besides the visible by changing the detectors.

What would the world look like to someone who could see into the ultraviolet and the infrared? In Alfred Bester's *The Stars My Destination* a woman can see heat, and it's like nothing she has ever witnessed—but Bester was just guessing.

The UV world is full of sharp contrasts, tending to black-and-white vision and fantastic patterns on flowers. But that is what we see on UV-sensitive photographic films, a special filter, so the world could actually look different from that.

In Frederik Pohl's 1975 novel *Man Plus,* the hero is an extensively modified cyborg with augmented eyesight. A nurse brings a bouquet of roses to his room soon after his new eyes are hooked up.

> Roger sat up and began again his investigation of the world around him. He studied the roses appraisingly. The great faceted eyes took in nearly an extra octave of radiation, which meant half a dozen colors Roger had never seen before from IR to UV; but he had no names for them, and the rainbow spectrum he had known all his life had extended itself to cover them all. But it was not quite true even to say that it seemed to be red; it was only a different quality of light that had associations of warmth and well-being.

We have already seen representations of IR vision in such films as *Predator,* but they are mostly guesswork. Certainly warm-blooded animals will light up against a dull background, at least at night. But as sensitivity increases beyond what we know now with IR goggles, the IR world will seem different. A whole new landscape of perception will open up, with colors for which no human language has words. The sun emits plenty of IR light during daylight hours with which to see, and at night infrared from nearby heat sources would light up the dark with subtle signals.

At the other end of the rainbow, what's beyond ultraviolet? Even more energetic waves, venturing into X-rays. Superman had X-ray vision, but that's clearly impossible for us. To see something with X-rays, there has to be an X-ray source illuminating the object. X-rays are so energetic, they destroy biological tissue. Luckily, our sun is a very weak source. The few solar X-rays are absorbed by molecules in the upper atmosphere—a good thing for life on the surface, but also of no use for seeing by.

Even Superman would be essentially blind here, because there would be nothing to see by. Unless, of course, he emitted his own X-rays like a flashlight, illuminating a target. This would give him plenty of information, but might very well kill the subject. (So much for the fear that an unscrupulous Superman would look under women's clothes! Offended modesty would be the least of their problems.) Superman would have to be immune to the lancing damage of X-rays, too, and that is a tall order for any molecule, much less the rather fragile ones that make up living tissue.

Imagine that we reverse direction, and go equally far into the lower frequencies, below the infrared. There lie the microwave frequencies that carry cell phone signals and high-data-rate transmissions for our communications systems. Even longer waves carry radio and television signals. There is not a lot of illumination in these ranges not provided by our technology—that is, the natural microwave world is largely a dim blur.

To see these waves, we would need very large eyes, because lower frequencies mean larger wavelengths. An eye that can sense a given wavelength must be several times the size of that wavelength, if it is to have any resolution at all. Imagine a microwave-seeing beast, with eyes the size of dinner plates. Obviously, these cumbersome orbs would have few evolutionary advantages over our smaller eyes—though this fact might explain some late-night TV monsters with bulging eyes and ferocious appetites.

Nature has exploited most of the electromagnetic spectrum it could, on our world. Apparently, the narrow channel of information we can see arose from what our atmosphere let through plentifully from our star. Electronics can extend this, letting us into the shadowy worlds of IR and UV, but not much beyond. Our window on the world opens only so far, and no more.

And what use could such a broader spectrum have? Without experience, we cannot say, but astronomy is a guide. By expanding their "window" into the radio and up into the X-ray, astronomers have learned much about stars and the filmy plasma between them. In a few decades radio astronomy alone gave us more information about the origin, nature, and destiny of the universe than millennia of philosophical reasoning.

The same might be true of our surroundings. Perhaps our world has information-saturated channels just beyond our vision. It will take careful, long studies to see if that proves so, but the potential rewards are vast.

A More Smelly World

The human nose contains millions of receptors, sensitive to a wide range of molecules that we interpret as smells. As good as they are, however, our noses don't compare to those of many other animals. Bloodhounds follow people's scent trails, pigs detect the buried fungus known as truffles, even our pet dogs and cats navigate in a world of smells we don't perceive. Scent molecules that our noses can't detect fill the air.

Unlike hearing aids, which have a long history, there have been no smelling aids—until now. A handful of researchers are working on artificial noses, mostly for industrial uses, like detecting spoilage in food-processing plants. With just a few dozen receptors, the devices have a long way to go before they can replace or even augment human noses. But they may see their first public uses in food packaging, where a sensor could detect the faint smell of bacteria long before the human nose could. Unlike a human nose's, the sensitivity of an e-nose is adjustable.

Digital smelling does not work the way biological smelling does. In the most common technology, e-nose recognition of an odor involves subtle changes in a thin sheet of carbon-impregnated plastic. The carbon particles absorb some of the odor molecules and swell up, causing changes in the electrical conductivity of the detector. Different odors are absorbed in different amounts, so each odor will have a different electrical "signature." The user of the handheld electronic nose has to "train" the device to detect the odors that are expected for a particular job. In a sense, this is not unlike training a truffle-sniffing pig to recognize the fungus's particular odor.

But animal noses work in a different way—through chemical bonding of an odor molecule to a specific receptor on a cell in the nose. Smell receptors are on the outside membrane of cells lining the nose. When a receptor meets an odor molecule, it binds to it, then sends a signal to the brain that is interpreted as a smell.

So in order for an individual to "smell" via the device, an artificial nose would have to be coupled directly to the brain, bypassing the nasal cells. Let's assume that can be done, though this technology awaits a better understanding of how the brain works, as well as compatible biomaterials for the implants. There are many smells we cannot detect, either because they are too faint, or because we don't have receptors for them. With an e-nose, we could sense a huge additional range of chemicals in our environment.

Conversely, we could train an e-nose *not* to detect a certain unpleasant odor (by not training it to recognize that odor). Combine the e-nose with nose filters, and the whole world could smell like rosebushes, no matter where you are.

A robot fitted with an e-nose could be trained to sniff out traces of explosives, drugs, hazardous industrial chemicals, or be used for quality control in food or cosmetic processing. Want each batch of perfume to smell the same? Compare their electrical signatures in the e-nose. Electronic devices are on the way to replace human testers for such products as blended scotch, perfume, wine, and cheese.

Possibly the e-nose of the future could help the more than two hundred thousand people each year who seek medical help for a deficient sense of smell. Without smell, most food is tasteless (as when a head cold temporarily knocks out the sensors in the nose). Because it is such a novel technology, and a sense of smell is not crucial, the e-nose lags behind cochlear implants. It will improve.

What we learn from the unfolding technology will probably eventually propagate into the broad market for people with normal senses. Devices first developed to aid the afflicted can improve the faltering abilities of the normal. It may be only a matter of time before we can retain our sensitive noses through our entire, ever-lengthening lives.

The inevitable next step will then enhance the sense of smell to unknown heights. Want to sense the forest as well as your dog? It will be possible. We know that people emit different scents as their body chemistry changes; the classic observation is that dogs can smell fear. Many people can sniff out sexual arousal, and laboratory tests show that these scents have a broad range of amplitudes and of taste. It might be possible to enhance one's sense of smell with a simple pill. Taken before going to a cocktail party, it could let us pick up faint signals of dislike, attraction, anger, or other emotional responses. There is some evidence that very social people—for example, individuals like Bill Clinton—can integrate scent, body language, facial expressions, timbre of voice, and other signals to anticipate and use social moments.

Very probably, if scent augmentation can be integrated with other perceptual skills, it will be popular. The future may indeed be more sensitive, if only technologically so.

Artificial Tongue

A related sense is taste, since most of what we think of as a food's flavor is actually smell. As we chew, volatile chemicals released from the food are picked up by smell sensors in the nose. Our taste-sensory cells are located in the small protrusions known as taste buds on the tongue. Despite all the tastes we can recognize, the sensory cells actually detect only a half dozen basic ones—sweet, bitter, salty, sour (acid), astringent, and umami (discovered by Japanese scientists)—each detected in a distinct region of the tongue. These detectors recognize different molecules by their shapes—the simple sugar glucose is a ring-shaped molecule—or the positive electric charge of table salt.

Umami is particularly striking, for there is no good Western word for it. It gives chicken soup its protein-rich flavor. It elicits our sense of full-bodied content in fish stock, cured meat, aged cheese, tomatoes, soy sauce, mushrooms and seaweed, even mother's milk. This sensory heft probably made us evolve a taste for umami because it promises useful content in proteins that otherwise would not have much appeal. Monosodium glutamate has umami. Salt, sugar, and umami send us primal signals about calories, proteins, and amino acids.

As happens with smell, receptor molecules on the surface of the sensory cells bind to appropriate chemicals coming in, and this triggers sensory neurons. That data is transmitted to other neurons in the brain, mingled with input from other taste buds and smell receptors in the nose, giving us the complex mix we recognize as flavor. The tip of our tongue picks up salty and sweet, the tongue sides best detect sour, while umami and bitter are best sensed at the back.

The e-tongue under development at the University of Texas, Austin, is not so divided; it works through a small piece of silicon indented with hundreds of tiny pits. In each pit are receptor molecules specialized to recognize and bind with a specific substance, like sugar, and measure how much of the substance is present. Biological tongues do this also, allowing us to detect from slightly sweet to intensely sweet. But whereas a biological taste bud can detect sweetness in many different chemicals, each sensory pit in the e-tongue can only detect a specific molecule. Depending on the intended use, this is either a limitation or a benefit.

For now, industrial uses are most likely—like flavor control in milk, or detecting spoilage in food. Applications would be anywhere taste is an indicator of a specific chemical, but especially when the chemical is dangerous, or where knowing the exact quantities of the chemicals is the goal.

Even here, the e-tongue can identify only what it's been programmed to find. This limits its use if, for example, your nosy neighbor wished to find out what secret ingredient your mother used in her meat loaf.

And what about augmenting our culinary sensitivities? Equipped with state-of-the-art electronic nose and tongue, could a flannel-tongued chili-and-beans lover be effortlessly transformed into a food connoisseur able to divine nuances of *boeuf bourguignon*? Would we become a world of ultimate foodies, savoring fine wines and aromatic dishes with the same full depth of perception as the finest naturally experienced gourmets?

Taste would still need training, allowing some to be snobs—probably an eternal need in our hierarchical species. Still, such a future seems possible, perhaps inevitable.

Nanodreams

The gaudy twentieth century was dominated by bigness—big bombs, big rockets, big wars, giant leaps for mankind—and perhaps the next century will be the territory of the tiny.

Biotech is already well afoot in our world, the stuff of both science fiction and stock options. We think of it in terms of drugs, of medical operations, and cures. Yet the finer one looks on the scale of biology, the more it looks mechanical in style. Consider the tiny beating hairs, flagella, that let bacteria swim through their fluid world. These work by an arrangement that looks much like a motor. Each proton extruded by the motor turns the assembly a small bit of a full rotation, along a helical axis. It is an elegant motorboat, engineering by evolution.

Biology operates on molecules, at scales of ten to a hundred nanometers (a billionth of a meter). Below that, from a few to ten nanometers, lie atoms.

Nanotechnology—a capability now only envisioned, applauded, and longed for—attacks the basic structure of matter at the nanometer scale, tinkering with atoms on a one-by-one basis. It vastly elaborates the themes chemistry and biology have wrought on brute mass. More intricate, it can promise much. How much it can deliver depends upon the details.

If one is able to replace individual atoms at will, one could make perfectly pure rods and gears of diamond (already five times as stiff as steel) that could be fifty times stronger. Gears, bearings, drive shafts—all the roles of the factory can play out on the stage that for now only enzymes enjoy, inside our cells.

For now, microgears and micromotors exist about a thousand times larger than true nanotech. There are already structures at the nanometer level, and some are remarkable. Pillars of orderly carbon atoms, arranged in a helix, make up "nanotubes" about ten carbon atoms across. They are a hundred times stronger than steel and conduct electricity, so they can become circuit elements. They have a huge surface area for their mass, and so could store electrical charge, making them into supercapacitors.

Ribbons of nanotube meshes could make light, protective vests for an energy-sufficient soldier, storing charge to drive his battlefield sensors and communications. Nanotube fibers could even be woven into artificial muscles which bunch and relax as charges are liberated from their helical tissues. Larger carbon fibers have already been made into sheets and films, with applications to light spacecraft.

By late 2000, this ever-shrinking frontier extended down to tiny machines made of nickel propellers grafted onto the shafts of four hundred biomolecular motors. Immersed in a bath of the basic energizer molecule of cells, ATP, five of the four hundred spun at speeds up to eight cycles per second. Film of the motion showed in several frames a dust mote caught up and flung forward. These were the first true nanomotors, microbe-size and drawing on the same biological energy source.

If somewhat smaller machines could self-assemble inside a living cell when needed, these "nanobots" could, for example, kill tumor cells by stimulating the synthesis of toxins. Later, other 'bots could round up the toxins, transport and release them directly inside cancer cells, leaving the tissue clear and healthy.

Life is already mediated by orchestrated biological nanomachines known as organelles coursing through cells. Seizing control at that level was until now the province of blunt forces of natural selection, operating through a long chain of chemical and biological links.

How small can these machines be? In principle, single atoms can serve as gear teeth, with single bonds between atoms providing the bearing for rotating rods. In the next few decades, making this happen may be only a matter of time and will.

Much excitement surrounds the possibility of descending to such scales, following ideas pioneered by physicist Richard Feynman, in his 1959 lecture "There's Plenty of Room at the Bottom." Later this view was elaborated upon and advocated by K. Eric Drexler in *Engines of Creation* (1986). Now some tentative steps toward the nanometer level are beginning.

Such control is tempting. As is the case with most bright promises, it is easy to see possibilities, but less simple to see what is probable. What should we expect?

Nanotech borders on biology, a vast field already rich in emotional issues and popular misconceptions. For example, many people, well versed in 1950s B movies, believe that radiation can mutate you into another life-form directly (not merely your descendants)—most probably, indeed, into some giant, ugly, hungry insect.

Not all fiction about nanotech or biotech is like this—there are good examples of firm thinking in Greg Bear's novel *Queen of Angels* and the anthology *Nanodreams* edited by Elton Elliot, and elsewhere.

All too often, though, in the hands of some scientists and science fiction writers, nanotech's promised abilities—building atom by atom for strength and purity, enabling dramatic new shapes and kinds of substances—lead to excess. We see stories about quantum, biomolecular brains for space robots, all set to conquer the stars. Or we witness miraculous overnight reshaping of our entire physical world—the final victory of Information over Mass, with even rocks and trees alive with information-intensive nanoforms that respond to their environment. Or about accelerated education of our young by nanorobots that coast through their brains, bringing nanoencyclopedias of knowledge disguised in a single mouthful of Kool-Aid.

Partly this is natural speculative outgassing. One can make at least one safe prediction: such wild dreams will dog nanotech. Genomics has a long experience with this. The late nineteenth-century ideas of eugenics led to bizarre experiments, and ultimately to the racial extinction policies of the Nazis, who systematically eliminated Jews, homosexuals, gypsies, and other "deviants."

As with genetics, the real difficulty in thinking about nanotech possibilities is that so little seems ruled out. Agog at the horizons, we neglect the foreground limitations—both physical and social.

Nanotech holds forth so much vague, speculative promise that writers can appear to be doing science fiction while, in fact, just daydreaming. Not only is the metaphorical net not up on this game of dream tennis, it isn't even visible. Drexler began serious nanotech discussions by bringing up the "gray goo" problem. What if the elementary builders introduced into an environment—"assemblers" that prowled, say, a bloodstream, searching for the right elements to combine into the desired end product—overrode their programming? They could then replicate themselves ad infinitum, eating everything in their path, like tiny omnivorous engines of destruction. This would destroy the body. More generally, assemblers might run wild in the natural world, leaving behind them nothing but manufactured goo. This is *Fantasia*'s sorcerer's apprentice with a vengeance. Sketching out the far reaches of the possible in his doctoral thesis, *Engines of Creation,* Drexler told enthralling stories of both nanoboom and nanodoom, and founded an entire field.

Ordinary adults can tell disciplined speculation from flights of fancy when they deal with something familiar and at hand. Nanotech is neither. Nature has made viruses, which are nanosized, but they cannot reproduce without coopting the mechanisms and resources of living cells. In billions of years, nature has yet to make autonomous nanobots of its own. This limitation suggests that there are some stumbling blocks ahead.

Nobody knows how to make an assembler, much less one that can eat anything and fend off competition to make gray goo. In biology, simple molecules can copy themselves. That does not mean that nanomachines at least ten times smaller can do what biological systems do routinely—ingest food, make energy, assemble proteins, and make more of themselves.

Worse, nanotech touches on the edge of quantum mechanical effects, an uncertain ground for engineering. Nothing in modern physics has been belabored more than the inherent uncertainties of the wave-particle duality and the like. People often take uncertainty as a free ticket to any implausibility, lofty flights of fancy leaving on the hour.

So nanotech will have to overcome several obvious basic design problems, as pointed out by Nobel laureate Richard Smalley. How can an assembler, a nanobot, not have impossibly "fat fingers"? Eight oxygen atoms in a row are a nanometer wide. How can an assembler of that size insert itself into a molecule to pluck out a single atom? Those other atoms will get in the way of any assembler.

Chemistry tells us that several atoms are linked together into a molecule by considerable energies, so the nanobot will need to rend them apart, using some other source of energy to do the work. It cannot simply harness its own binding energies, or it would itself come apart. Instead, there must be nearby chemical resources that the nanobot can harvest. Electricity won't work in wet surroundings, so the resource will have to be a ready chemical store.

Finally, how to let go? Atoms must stick to the assembler to be moved, so some chemical energy must be used to make the connection. But then the assembler must let them go. How? Inverting the chemical reaction demands some other agency, adding complexity to the process.

These are indeed stumbling blocks. Engineering may leap them in time. Add to these the problem that for now, at least, nanotech engineering is carried out at a few degrees above absolute zero temperature. Atoms scarcely move at those temperatures, making them at least easier to hold on to.

But in the real, ordinary world at room temperature, where workaday nanobots would need to work, atoms vibrate ten times faster than at cryogenic levels. That makes an assembler's task very hard indeed, especially since it shares this incessant vibration. A nanobot will have to be both tiny and fast.

That is not to say that nanotech is impossible. Drexler's vision still may bear fruit. But dreams can be set aside for a while. Developing a discipline demands discipline. Dreaming is not enough.

One point we *do* know must operate in nanotech's development: nothing happens in a vacuum. The explosion of biotech, just one or two orders of magnitude above the nanotech scale, will deeply shape what comes from nanotech. By the time nanomanipulation becomes practical, decades of biological engineering will have given us a profusion of "miracle" products and methods.

These innovations will have stretched the average life span a bit, and enriched our later years. It is easier to implement biotech because the tiny "programs" built into life-forms have been written for us by nature, and tested in her remorseless lab. Not so nanotech, which probably will be used to attack problems even more fundamental than, say, cancer—disorders we may not even know of now.

The transition from biotech to nanotech will be gradual. Above the tiny nanoscale, the "biologic" of events is flexible, compared with mechanical devices. Below it, functions are increasingly more machinelike. The ultimate limit would be the nanotech dream of arranging atoms precisely, as when a team at IBM spelled out the company initials on a low-temperature substrate. But widespread application of such methods lies probably decades away.

The future will be vastly changed by directed biology, before nanotech comes fully on stage. Already some are thinking beyond that stage, to a future that may greatly extend our life spans. But one must reach that era, which may lie a century or more away. We mortal, aging inhabitants of the early twenty-first century will not last that long. What to do?

Consider cryonics. This freezing of the recently dead, to be repaired and revived when technology allows, is a seasoned science fictional idea, with many advocates in the present laboring to make it happen. It is perhaps the best case where nanotechnology will augment our bodies in ways most do not even consider. Nanotech is probably essential to cryonics, which has a determined subculture. These believers, who cluster around a few companies like

Alcor, trust that an investment now can pay enormous dividends in the distant future.

While nanotech promises to treat disease and decay at the most subtle, fundamental level by reordering atoms, cryonics invokes nanotech to make the ultimate promise, extending beyond what we regard as death itself. Cryonicists support an expanded definition of death—that to be truly dead means that one's identity has been so destroyed that recovery of it lies beyond the limits of any conceivable *future* technology.

Neil R. Jones invented the idea in a science fiction story in the 1931 *Amazing Stories,* inspiring Dr. Robert Ettinger to propose the idea eventually in detail with *The Prospect of Immortality* in 1962. It has since been explored in Clifford Simak's *Why Call Them Back from Heaven?* (1967), Fred Pohl's *The Age of the Pussyfoot* (1965), and in innumerable space flight stories (such as *2001: A Space Odyssey*) that use cryonics for long-term storage of the crew. Author Larry Niven coined "corpsicle" to describe such "deanimated" folk.

In large measure because of such dreamers, cryonics is also real, right now. Ted Williams was frozen in 2002, and Timothy Leary had made arrangements to be a few years earlier, only to change his mind only days before his death. About fifty people now lie in liquid nitrogen baths, awaiting resurrection by means that must involve operations below the biotechnical.

Repairing frozen brain cells that have been cross-slashed by shear stresses, in their descent to around 77 degrees absolute, then reheated—well, *that* is a job nothing in biology has ever dealt with.

Such tools imply a kind of medical technology that can have vast social implications, indeed. The example of cryonics shows just how far augmentation might go—to the very edge of the grave, and then beyond.

Naturals and Mechanicals

Here is where the future peels away from the foreseeable. Nanotech at the cryonics stage will drive large qualitative changes in our world, and undoubtedly in our worldviews, that we simply cannot anticipate in any detail. From here, it tends to look like magic.

Suppose the next century is primarily driven by biotech, with nanotech coming along as a handmaiden. Do we have to fear *another* radical shift in ideas, with nanotech?

Biotech and its implications—androids, cyborgs, and bionic beings—look at first glance superpowerful, but remember, evolution is basically a kludge. Organisms are built atop an edifice of earlier adaptations. The long zigzag evolutionary path often can't take the best, cleanest design route.

Consider our eyes, such marvels. Yet the retina of the vertebrate eye appears to be "installed" backwards. At the back of the retina lie the light-sensitive cells, so that light must pass through intervening nerve circuitry, getting weakened. There is a blind spot where the optic nerve pokes through the optical layer.

Apparently, that was how the vertebrate eye first developed, among creatures who could barely tell darkness from light. Nature built on that, developing lenses that could focus, better retinas. And nature found different paths altogether; the octopus eye evolved from different origins, and has none of these drawbacks.

Could *we* do better? Consider first not implanting better sensors, as we have described already, but retooling through biology itself. A long series of mutations could eventually switch our light-receiving cells to the front, and that would be of some small help. But the cost in rearranging would be paid by the intermediate stages, a tangle that might function more poorly than the original design. Nobody is going to try this on people until extensive tests work on animals.

So these halfway steps would be selected out by evolutionary pressure. The rival, patched-up job works fairly well, and nature stops there. It works with what it has. We dreaming vertebrates are makeshift constructions, built by random time without foresight. There is a strange beauty in that, but some cost. We work well enough to get along, not perfectly. Ask anyone whose mostly useless appendix has burst.

The flip side of biology's deft engineering marvels is its kludgy nature, and its interest in its own preservation. We are part of biology, integrated into a system that cares mostly about the next generation—that is why sexual desire comes upon us so soon, before true maturity; the world is going to need more people. Biology in the large is seldom our submissive servant, except when it can benefit our successful reproduction. In the long run, the biosphere favors no single species.

The differences between nanotech and biotech lie in *style*. Of course functions can blend as we change scales, but there is a distinction in modes.

Cells get their energy by diffusion of gases and molecules. Nanobots may be driven by electrical currents on fixed circuits. Cells contain and moderate with spongy membranes; nanoengines must have specific geometries, with little slack allowed. Natural things grow "organically," with parts adjusting to one another; nanobuilders must stack together identical units, like Tinkertoys.

The Natural style versus the Mechanical style will be the essential battleground of tiny technology—and of cyborgs and androids. Mechanicals we must design from scratch. Naturals will and have been force-evolved; their talents we get for free. Each style will have its uses.

Naturals can make things quickly, easily, including copies of themselves—reproduction. They do this by having what Drexler terms "selective stickiness"—the matching of complementary patterns when large molecules like proteins collide. If they fit, they stick. Thermal agitation makes them smack into each other many millions of times a second, letting the stickiness work to mate the right molecules.

Naturals build, and as time goes on, they build better, through evolution. In Naturals, genes diffuse through the population, meeting each other in myriad combinations. Minor facets of our faces change so much from one person to the next that we can tell all our friends apart at a glance (except for identical twins).

These genes collide in the population, making evolutionary change far more rapid because they can spread through the species, getting tried out in many combinations. Eventually, some do far better, and spread to everyone in later generations.

This diffusion mechanism makes sexually reproduced Naturals change constantly. Total Mechanicals—robots of any size, down to nanotech—have no need of such; they are designed. There is no point in building into nanomachines the array of special talents needed to make them evolve—in fact, it's a hindrance. It could become a danger, too.

We don't want nanobots that adapt to the random forces of their environment, taking off on some unknown selection vector. We want them to *do their job*. And *only* their job.

So nanotech must use the Mechanical virtues: rigid, geometric structures; positional assembly of parts; clear channels of transport for energy, information, and materials. Mechanicals should not copy Naturals, especially in aping the ability to evolve.

This simple distinction should lessen many calls of alarm about such invisible, powerful agents. They can't escape into the biosphere and wreck it. Their style and elements are fundamentally alien to our familiar Naturals.

Nanobots' real problem will be to *survive* in their working environment, including our bodies. Imagine what your immune system will want to do to an invading band of unsuspecting nanobots, fresh off the farm.

In fact, their first generation will probably have to live in odd chemical soups, energy rich (like, say, hydrogen peroxide or even ozone) and free of Natural predators. Any escaping from their chemical cloister will probably be eaten—though they might get spat right back out, too, as indigestible.

The "gray goo" problem of nanotech, in which ugly messes consume beautiful flora and fauna, need not occur, precisely because the goo will be gray, mechanical, and vulnerable. Nanomachines need not have the rugged, hearty defenses that are the down payment for anything that seeks to use sunlight, water, and air to propagate itself. Gray goo will get eaten by green goo—maybe by an ordinary slime mold, which has four billion years of survival skills and appetite built in.

So nanotech will not be able to exponentially push its numbers, unless we deliberately design it that way, taking great trouble to do so. Accidental runaway is quite unlikely. Malicious nanobots made to bring havoc, though, through special talents—say, replacing all the carbon in your body with nitrogen—could be a catastrophe.

What uses we make of machines at the atomic level will depend utterly on the unforeseeable tools we'll have at the molecular level. That is why thinking about nanotech is undoubtedly fun, but perhaps largely futile, for now. Certainly such notions must be constrained by knowing how very much biology can do, and will do, long before we reach that last frontier of the very, very small.

On that nanoscale, probably the first applications will be implantable sensors that can sniff out the early, small signature molecules of cancer. Those sentinels will be large and rugged at first, and will get smaller and less perturbing in time. Probably among the first patients will be some who balk at having small artificial additions in their organs or bloodstreams. That will be the first encounter most have with technology they will probably barely be able to see.

Still, most people will care little whether the miracles ocurring in their bodies are driven by biologically derived molecules, or by artificial nanobots. They will marvel for a while that such capabilities occur, and then accept them as "natural" outcomes of a technology that becomes routine. Once again, we will redefine ourselves as ordinary, when in fact an ancient Egyptian would find us wondrous.

Such is the nature of progress—to make itself gradually invisible.

CHAPTER **3**

Chips, Brains, and Minds

*By adding chips to ourselves, can we finally conquer
fear of math? Become Einsteins? Or maybe just find our car keys?
And what of our minds? Our selves?*

Our Computers/Our Selves

Are We Really Computers?

You probably own a computer of some kind, if only a hand calculator. Owning it seems natural, convenient, unremarkable. Never, for example, do you think of yourself as a slave owner.

But someday you might be—in a way. If our silicon conveniences get more complex, can they pass over a fuzzy boundary into selfhood?

That depends on how you think of them, and of us—which again depends on how you think, literally. With the spread of personal computers, a common analogy has crept into the way we think of ourselves, with implications that we don't truly understand.

Much science fiction, and a lot of everyday talk, assumes that we are much like computers. Simply put, the analogy goes like this: Brains resemble "wet-ware" computers, and minds are like programs. We, our precious Selves, are programs running on our brain-computers.

Is this reasonable? Here we enter the swamp of metaphilosophy, though with a life raft of experience for consolation.

Start from the physical facts. Computers are tiny circuits etched in silicon chips. They have been made of other things, though, sometimes as simple as Tinkertoy mechanical parts. These oddities show that programs can run on several different types of hardware, displaying properties independent of the hard "substrate." So it seems plausible, to some, that our brains might be just another substrate, although rather slippery and less reliable.

We're intelligent, by (our own) definition. Computers running on hardware made of silicon are getting faster, more powerful, increasing their total memory by the year. Does this mean that they will eventually become intelligent?

There is a vast difference between intelligence and raw processing power. Within two or three decades, computers will probably reach the human level of data processing—about ten "teraops," where a teraop is a thousand billion operations, or bits of information (basically, a choice between a zero or a one in a particular slot) per second. This huge number, 10,000,000,000,000, is the number of nerve cell synapses available to fire at any time, not the number of ideas or sensations you register. Sensations are whole constellations of information, just as the brush of a breeze is a multitude of molecules caressing your cheek. In conceptual space, we can take in consciously about one new piece of information per second. We're slow learners.

Still, let's call that number, ten trillion bits, a "Library," because the Library of Congress holds approximately that much stored in books. (The *Encyclopaedia Britannica* holds about a tenth of a Library, while our DNA carries a mere thousandth of a Library—though of course we have one in every cell. Nature likes redundancy.)

By the late 1990s, our data-collecting satellites that monitor the environment, carrying out NASA's Mission to Planet Earth, sent down to us about a Library of bits *per day*. Human brains process that fast already, mostly to do body maintenance.

We can hold in memory roughly a Library's worth of bits—far

more than, say, an alligator, which can remember at most about ten billion bits, a thousandth of our capacity. That's enough to get the alligator around his mucky world, but it does not qualify him to vote, attend Harvard, or speculate on the meaning of Self.

Computer speeds have increased hugely over these last few decades. Mind-power, though, is another matter, almost completely unconnected with speed. If we knew how to write a program as smart as you, we would not need to run it on a fast computer unless we wanted it to respond as quickly as you. Instead, one could take your Self Program 1.0 and run it on a laptop. With 40 megabits of memory, conceivably this machine could run "you"—but *very* slowly, and without some of your memories. It would behave like a very dumb, sleepy "you."

So speed isn't the essential here. To see if Self Program 1.0 is possible, we have to look deeper than counting bits and teraops.

Seeming Human

At a cocktail party you probably assess others' intelligence without thinking about it. That's the Turing test—people pass it every day, just by convincing you that there's some Self in there talking back. We assume people on the other end of a telephone conversation, for example, are not really "expert systems" designed to fool us. (Imagine a few decades from now, when this will be a real possibility. Already some answering machines seem smarter than the secretaries they replaced, or at least certainly more reliable.)

The Turing test is always limited; you don't ask an infinite string of questions to be quite sure you're talking to a true Self; you haven't the time. Even without it, however, we intuitively know that computers that can play chess or even diagnose diseases aren't candidates for their own Bill of Rights. Reliable, sure—just as a light switch is.

Computers do some difficult things, like balancing accounts or adding numbers or selecting from a menu of choices, very well. Other skills, natural to us, they struggle with—picking up a glass of water, say, or navigating around furniture. That is because they do not know how to make a working model of the world. One of the consolations of the Coming of the Computers is that in many functions lawyers and doctors can be replaced, but cooks and gardeners will remain human.

But merely looking at number-crunching capacity in a brain or a silicon computer still isn't enough. We have electronic data systems like ARPANET, Internet, BITNET, and so forth that weave webs of information exchange around the entire planet. With many millions of users, they have a computing power roughly like that of your brain, but nobody would confuse them with you. Though their cheerleaders believe these nets represent something qualitatively different in our culture, they are basically sophisticated mailboxes. Merely capitalizing the word "Matrix" does not make the nets self-aware entities, and adding more processing units will not help. They are passive, dumb.

What even "expert" programs miss is *complexity*—that still mysterious property that emerges from complicated systems as the details of the process blur into the background. In our brains, the snap of synapses lies behind a quick quip, but we don't sense those neural events. They're tiny.

Instead, the mind is better seen as a system of systems, a point of view Marvin Minsky advanced in his groundbreaking *The Society of Mind*. The characteristic of such complexity is that new things emerge from old. Fresh ideas bubble up to us, somehow concocted from the old, fixed information in our memories. How? Nobody knows—yet.

The complexity of Mind must be qualitatively different. We get an idea of how different that must be by looking at our picture of computers.

A laptop can run many different programs, but not all. It has a relatively small capacity, and many programs are written in a language unintelligible to the laptop. An IBM-type laptop won't run Apple-type programs, though one can translate between them already. But such facets are like the division of Christianity into Catholic and Protestant, and are bound to fade. In principle, any computer could run any program if it had the room.

Brains aren't like that. Our Selves consist of our hardwiring, the hard-won patterns we've grown. They are not "coded" into our brains, like sentences written on this previously blank page you're reading. Your mind could not run on another brain. A vast number of "mind-transfer" stories skate over this simple but profound point, to their loss. Not that mind transfer is forever impossible, though. The real story to be told about it is how difficult it will be, and what it may imply.

Sticking a wire into your head and jetting through computer spaces, as in *The Matrix* and its sequel films, is implausible. Computers are digital, sending short pulses of electricity at high speed, strings of zeros and ones that are then read as numbers or words. Brains are partially analog devices, like a mercury thermometer, which registers temperature by the height of a silvery column. These methods don't interact well.

A neuron can generate about a hundred signals in a second. Today's electronic switches (in chips) switch a hundred *billion* times per second. Obviously, there's a mismatch. Further, chips are made quickly and cheaply in factories. Neurons have to build themselves from the inside out, growing as a baby ages. This process leads to profoundly different rates of changing, learning, and forgetting.

When John von Neumann invented the modern computer, he was being quite practical. He had a hardwired computer, ENIAC, that connected given tubes and switches ("memory registers") with actual copper wires. To reprogram, von Neumann had to move the wires. He tired of this, and so invented the idea of programming. (This story echoes Robert A. Heinlein's observation that all real progress comes from laziness.)

Von Neumann designed new data memory registers so that they not only held passive information, like old data and more recent results, but could also retain instructions for how to flip electric switches. These switches made new connections to the existing wires in ENIAC. Those instructions were the first program. They did the rerouting of electrical current, instead of some poor technician with a screwdriver.

That is a deep difference between ourselves and our computers.

Our brains are like von Neumann's old ENIAC. To make and keep a memory, we form new connections between existing neurons. Those neurons are not simple switches; instead, they connect with a few thousand other neurons. Once made, connections last— they are kept "up and running" by the same ebbs and flows that keep the rest of your body going. That's why we take time to learn new things—the brain has to knock down some connections and rebuild them. Thinking hard, our brains can burn up to 40 percent of our calories. You really can tire from thinking, and about half of our heat loss occurs from the neck up.

The quality and kind of these many-neuron connections determine the kind of intelligence we have, the style of our Selves. The kind of connection is also task-specific among living things. Other animals process their data—that is, experience their surroundings— in ways we cannot understand or exactly duplicate. Bats see with sound waves, dolphins with sonar—and their brains work this information over differently, yielding a differently perceived world. Even the augmented senses of smell, sight, and hearing we discussed earlier will not give us the world as sensed by animals, but rather a different cosmology of senses, probably new, not a copy of that in any creature.

So your memory of your grandmother is not "coded" in the way that your airline reservation resides in a computer. Her image is hardwired, laid down in synaptic strata. The twinkle in your grandmother's eyes, so easy to call up, may be reached by different routes in your memory, too. You might recall the sled she gave you that distant winter, and see the twinkle. Or you might remember the sled's name—*Rosebud,* say—and reach the twinkle that way.

We naturally have something like holographic memory, with important information reachable by processes we sometimes can't even know. That's why you can often recall a friend's name by *not* thinking about him for a few moments. Your subconscious has gone on, rummaging through associations (other friends, meetings, "photos" kept in storage, like your grandmother's twinkle). Then the name pops into your head, retrieved by a roundabout route.

Smells often call up distant memories, apparently because the smell perceptors in the brain are near, in the neuron-connection sense, to the sites of long-term memory. For many, the smell of an approaching rainstorm evokes childhood. For others, a pop song from high school—say, Dylan's "Like a Rolling Stone," voted in 2004 by *Rolling Stone* magazine the best song of all time—casts them back into earlier feelings.

Minds as Webs

A better picture of our brains is really the spiderweb model, described in Minsky's book, in which distant strands tremble if you shake it in the right way. (Seeing the reflections from a sharp blue sea may recall that twinkle; or maybe a trip through your old school yard will, for reasons you can't readily call up. Again, we are mysterious to ourselves.)

To dissolve a strong memory takes wholesale destruction of many neuronal connections. You'll never forget your grandmother, but you will forget last Tuesday's lunch—in fact, probably already have. Rewiring takes time. Most of the day's events, though, you don't need to keep. They're repetitious, and would quickly take up all the room you have if you tried to hang on to them.

Of course, the "you" here isn't the conscious, voluntary you. It's something we crudely term the subconscious—to Freud, the unconscious. It works without your knowing it does, though at a far higher level than the operations that keep you running—digestion, breathing, heartbeat—down in the medulla, a knot at the join between your brain and the central nervous system.

When we sleep, our subconscious throws away most of the memories of the previous day, cleaning house and dusting off to make more room you'll need later. So you can't remember immediately what you had for breakfast twenty-four hours ago, and have to reconstruct it from hints. ("I usually have cereal . . . yes, and there were eggs, too. . . .")

This subconscious editing and garbage disposal are essential, because we use all of our memory fairly often, despite claims that we use only a fraction of our brains. Unused material is thrown away, unless it's powerful stuff, deeply implanted with many connecting routes (like your grandmother's smile, or your first date).

So in this way our brains are very different from our silicon servants. But there is a deeper level, requiring much more knowledge about ourselves than we now have, where the brain/computer analogy might work. Silicon computers actually rewire themselves, too.

My laptop running a new program differs internally from the same laptop last week. Programs set and reset tiny switches in the central processing unit, millions of times a second. The computer is ever-changing. Programs rewire computers in a flexible way. We can't see this, of course, but the computing "brain" alters constantly as a program runs.

So at the very minute level, our brains lie in a continuum with computers. In a meaningful way, a computer in a given state (with particular settings of its myriad switches) can run only one program—the one that sets those switches.

But it will take a long, long time before we understand brains well enough to make a computer program emulate a brain. Particularly since there is a hierarchy in all this that we can only dimly glimpse.

Conscious Computers?

To bring a computer to consciousness, one must first know what makes us conscious. We aren't remotely close to understanding that, so we concentrate instead on the minutiae of the problem. Some take the building-from-the-ground-up approach, and begin with the simple connections.

Switches are the underlying, simple building blocks. Can stacking layers of systems of switches eventually build to Mind? Where would the transition occur? We do not know, and no simple extension of machine capability seems likely to tell us. In Robert Heinlein's novel *The Moon Is a Harsh Mistress,* stacking up sheer processing power led to a sentient computer named Mike. Maybe the transition can occur like that, but we can scarcely count on such miracles.

We do understand that no given neuron or switch is essential to the Mind, and indeed our brains deal constantly with the quiet death of cells. What matters is the interactions a neuron has with tens of thousands of other neurons. When it dies, others take up the slack, embedding memory in other, fresh connections. Evolution favored this, since redundancy of function is a good safeguard against forgetting something vital. Cells come and go, but your grandmother remains, beaming at you in "wet-ware" memory.

Most "cybernetic philosophers" feel that mere information processing requires no fundamental complexity or intelligence. Making *new* information does—so Mind, with the capital letter, stands for Mystery, too—and for the fact that we do not experience how our own minds work. Though we live inside this marvelous machine, we do not have a clue about where the transition from a vast field of fast switches merges into a thing that knows itself to be alive.

Intelligence can't be the product of a simple rule-based system, such as expert systems that diagnose problems. It could mimic a human for a while, but ultimately would fail the Turing test. Only a rule-based program that had so many rules that the individual rules got lost in the noise would have a chance—and that would verge upon the swampy province where Mind emerges.

"We will be software, not hardware," Raymond Kurzweil predicts, but this assertion demands that we be reducible to a program. This need not be an insult to us. Programs are bits of technology that do jobs faster than we can, just as cars move faster than we can. A pocket calculator can beat us at arithmetic, but it is not intelligent, nor does it prove new theorems.

An advanced program, a simulation of the world, is not a duplication or re-creation of the world—it is a model. A map is not the territory. A simulation of your heartbeat does not beat itself, it rearranges meaningless symbols to represent the beat.

As well, computers simulate processes that can be described precisely. (That does not necessarily mean their outputs or methods are always precise; there is a field of "fuzzy logic" to deal with this uncertain aspect.) Life cannot be so described, except in stylized arenas like the chessboard. There, IBM's Deep Blue program beat world champion Garry Kasparov.

This human defeat did not stop people from playing and enjoying chess games. It merely told us that the absolutely well-defined world of chess can be mastered by a machine that knows nothing of that world. Deep Blue does not know what pawns are, only what symbol represents them. Nobody at IBM proposed Deep Blue for citizenship after Kasparov's defeat. It will be a very long wait before that problem arises, perhaps when we have a full understanding of how we manage consciousness.

For now, a circuit that plays a game is no more relevant to our self-respect than would be a steel football-playing robot that nobody could stop with a tackle. Robo–running backs would be a one-time-only oddity. Then sports fans would forget about them, because they are irrelevant to the true concerns of football, a game that for many satisfies deep needs—pursuit of opponents across terrain for a goal, just like soccer and many other athletic contests.

Perhaps in retrospect some degree of brain–computer resemblance is unsurprising, because we have created computers using our brains. Some aping of method is probably inevitable, though perhaps unconscious.

After all, there are other qualitative resemblances. Both brains and computers have fibrous structures (wires versus nerves), they use a traffic of signals following well-defined routes (not just broadcasting at large, as some electrochemical reactions do), and use pulses of electricity to carry out logical operations (changing local states).

Indeed, the recent land-rush euphoria in computing circles for "parallel processing"—running programs simultaneously, which then merge to solve a problem—echoes our own brains' method of dealing with several tasks at the same time. Our three-pound package of slow and slimy parts organizes in a phalanx of parallel processors, so that indeed, we can walk and chew gum at the same time—and think about the Self, in the bargain.

Uploading Our Selves

Can computers hold a "brain emulator" within themselves?

Duplicating all the neuronal connections is in principle possible, in the sense that we could run the brain emulator as a program on a vast computer, which in turn runs a metaprogram (Self, Mind) on a conceptual layer above the emulator. This is really only one grand program, of course, but it is easier to think of it as two-layered: The first floor imitates the wiring of our brains. One flight up, a program uses the information lodged in that wiring pattern to synthesize a sense of Self.

It's only one more step, then, to invent the idea of transferring our Selves to computers—termed "uploading." Charles Platt's *The Silicon Man* deals with the gritty feel this might have. Its plot armature is the central motivation: escaping death.

When might this be possible? Hans Moravec, in *Mind Children*, estimates that "human equivalence" of computing power in a supercomputer (a linear development of the Cray class of the early 1990s) will be possible around 2010. If personal computers keep advancing as they have before, they could run humanlike programs around 2030. Then your laptop could become your friend—or your slave, depending on your point of view.

How? The mere theoretical capability is no true guide. Despite the ease science fictional characters have interfacing with computers, there are all the decided conceptual problems discussed above, plus problems with humdrum technical hassles. The frequency of brain waves is a million times lower than that of computers, and their electromagnetic bandwidths are vastly different. Sensing the workings of a computer by plugging a wire into your brain would be like trying to take a drink from a firehose—most likely, you'll get nothing but a sore mouth.

Then there are the grungy details of how to extract the information from your head. In principle we could do this without knowing in detail how the brain works. Instead, we could use the principles of copying software, to recognize neurons and then replace all the functions of each neuron with a computer simulation.

Neurons hold your Self, encoded in their myriad connections. It's not enough to know the location and type of neuron; one must also see how each one responds and sends electrical signals, how it is affected by its chemical environment, and so on. You think differently when your adrenal glands have been squirting into your bloodstream—as anyone with a temper knows.

So we would need to assess the functioning of each neuron under different conditions. This might demand that a neurosurgeon insert microscopic machines that can sit atop a layer of the brain, registering how you think while you're subjected to a number of influences, probed by stimuli, and performing some thinking tasks. A sheet of these sensors covers the crown of your brain as the process begins, building a three-dimensional map of a thin layer of your brain cells. Added to a general map of human neural structure, the surgeon writes a program that models the way your brain layer works, all the idiosyncratic ways you think.

A working model can then be sharpened by you and the surgeon, by comparing its output signals to those you emit, given the same stimulus. Flash by neuronal flash, this computer model is made to exactly echo yours. Once they correspond, a bit of your Self resides in a computer. The trouble is, that tactic works fine for the top layer—but what then? To reach the next layer down, the surgeon's easiest path is simply to shave away your brain, or render it dormant in some fashion. It seems quite plausible that destroying that layer might be the only course for quite a long time in the evolution of neurosurgery. Your brain, to be fully read, must die.

You end up with an excavated skull, perhaps without even interrupting your train of thought or perceiving any pain. (Luckily, the brain has no pain perceptors in its spaghetti snarls of nerves.)

Not a voyage for the squeamish. Obviously, the material Self is gone. Your represented Self remains, in silicon. It says so, right here in the contract.

Moravec suggests a halfway house for this journey into digital immortality. Our brains are in fact already a house divided. The left half controls the body's right side, right hemisphere managing the left. They also specialize in technical tasks, conferring back and forth with each other, connected by thick bundles of nerve fibers, the corpus callosum. Severing these fibers doesn't shut down the brain; rather, each side proves to be an independent, intelligent, fully conscious Self.

Suppose, half a century or so from now, a surgeon cuts that nerve highway between your hemispheres. A computer can eavesdrop on the data flow between the two hemispheres, then pass it on, keeping your bisected Selves in touch with each other. The computer cooks up a model of how you operate. As your brain ages, losing cells and functions, your keen wit slowly blunting, then the computer program can insinuate itself, keeping up your mental crafts. Eventually, your brain loses so many cells, and perhaps succumbs to various diseases of degeneration, that it dies. But the program remains, suitable for installing permanently on a computer—without any sense that your Self has ebbed.

Is this you? That is the essence of a deep identity problem—continuity. When we sleep, the subconscious remains active, ensuring continuity at a broad level. No one wakes up thinking he is a new person. Patients brain-cooled until their brain waves lapse can later revive with their sense of self intact. But are they "really" the same?

They awaken into a world unchanged. That would not be so if they were revived, say, ten years later, or in India. That sort of shift would surely disorient them. In many science fiction stories, characters struggle to have copies of themselves stored, or do indeed awaken as copies of some dead original, and go on about their business. It's all quite blissful.

And implausible. There is a huge difference between the *inner* sense of Self and outer appearances. They are not the same. Failure to see this distinction leads to confusions.

Matter transmitters that destroy people and recreate them elsewhere do not satisfy the continuity condition, because the original knows it dies. One can disguise this by making both death and reconstitution instantaneous, of course. The *Star Trek* crew never flinch at the prospect of being disintegrated when they beam up, because by sleight of hand the *Trek* technology does not actually harm anything while "reading" the person. Instead, the actual atoms of Spock, say, are sent winging their way back to the *Enterprise,* and then Spock gets rebuilt. This is yet another step beyond the destroy-and-rebuild scenario we have invoked. It seems proportionally more implausible—beams of atoms are notoriously harder to transmit than beams of electromagnetic waves.

So fans of uploading must promise continuity, by showing that the underlying structure (the brain) is continuous, even if consciousness and brain activity are not. Matter transmission, on the other hand, has no inherent continuity of substrate or "software."

At sufficient distance or time, there is no way to tell if a transmitted person is the same as the original, since you cannot compare it to the original, and you can't count on those who knew the original to sharply remember him. One must interact with a copy to see if it carries the true deep content of an original.

The fallback answer is to keep the Self wide awake while the transition goes on, so that it knows it is intact. Those are the only terms one should accept in any uploading scheme. Otherwise, there's a real chance that at the deep level, you're merely entering into a suicide pact.

All this discussion points to a glaring deficiency in our understanding of our selves: we have no deep understanding of how we think, of what the Self is. Some say it is a mere convenient fiction—in computer lingo an "operating system"—that lets us manage matters.

There is probably no more difficult problem in science than the blend of experiment and philosophy that attempts to confront this problem of defining Self. When you get an exceptionally smart home computer, a few decades hence, you may find such discussions becoming real, everyday issues. How will we think about them— and about ourselves?

Often we are most involved, less self-conscious, when we are in the grip of strong emotion. Is that a clue? Are emotions necessary to minds such as ours—and will they be equally so for all manner of mentally augmented people? Or robots? This consideration is a crucial feature of any discernible future for these broad frontiers.

Reasoning About Emotions

Your brain has about a hundred billion neurons, roughly the same as the number of stars in the Milky Way. Your liver has about a hundred million cells, but a thousand livers strung together do not add up to that mysterious sense of Self. What distinguishes the brain is its organization.

"You"—the thing that Descartes thought unquestionable when he said, "I think; therefore, I am"—is a rather abstract entity. Somehow it emerges from three pounds of wrinkled jelly that with a few applied volts can ponder and feel, invent calculus, and know joy. That Self apparently resides mostly in the highly folded surface of your brain—the cortex, from the Latin for "bark."

The cortex isn't a woody protective coating—it's apparently the boss. Yet it's only about five millimeters thick, and even with the convoluted folds covers about the same area as an office desk.

Think of your Self, spread out like a big beach towel. Storms and waves of electrical impulses race across it. Slow oscillations of about forty cycles per second seem to supply the familiar, comforting sway of predictable "weather," much as the day–night cycle does for the body. Fitful breezes trace the sweep of momentary sensations. A passing spiky swirl tells of an interesting conversation. And sometimes—less frequently, as we age—great hurricanes blare through, seizing us: the emotions.

Here we connect most profoundly with the body. Emotions reach beyond the computational abstractions of the cortex. They stimulate hormones, accelerate your pulse, quicken your breathing—which in turn affects your mind, too. It's an illusion to merely think of our minds as cool calculators.

Einstein once remarked that he didn't see his theories in mathematical language, much less in ordinary prose. He *felt* them. Kinesthetic sensations had to come together in a satisfying way, so that he experienced the *rightness* of the idea.

As a teenager, he wondered what would happen if he moved near the speed of light, while holding a mirror in front of himself. When his speed approached that of light, would he suddenly see his own reflection wink out? If so, that meant that light had been bested by his motion. It wasn't a crazy idea. That is indeed what happens to sound waves when an aircraft exceeds Mach 1, moving faster than the speed of sound. Above Mach 1, sound cannot catch up with the airplane, and it forms a shock wave in air, which we hear as a sonic boom—all the energy stacked up in a wave's thin leading edge.

Einstein *felt* that this imaginary experiment was wrong. It took him another decade to see why, when he devised the special theory of relativity. He simply invoked the constancy of light's speed as the opening postulate to the theory, not justifying it except by the predictions implied—which turned out to be true, years later. He did this out of his kinesthetic sense.

This anecdote suggests that thinking about our minds as calculators, neatly divorced from the rub of our world and doing computations in a void, misses an important lesson of evolution. We have large, highly developed sensory and motor portions in our brains. Our nervous system is very specific. That complexity is why cavalier claims about linking computers to brains are mostly just daydreams.

We are highly adapted creatures. Our neural nets mirror our survival strategies, right down to the basics that link us so strongly to each other. While plenty of biologists stress how large our brains are, few mention that we have the largest genitalia of all the primates, in ratio with our body weight, and that they capture a disproportionate fraction of our nervous system, too. The sensory and motor parts of our brains have about a million times the effective computational power of our conscious minds. We can walk the walk *and* talk the talk.

Novices at a craft can get by with "book learning"—that is, computational schemes worked up without experience (but based on our understanding of how the world works). To become expert, you need hands-on experience. We "map" our body sensations into representations in our minds. "Flying by the seat of your pants" means that you let the body do a lot of your thinking for you. When you get into your car at the end of your working day, unless you explicitly remember that you have to stop by to pick up a gallon of milk, the sight of familiar avenues will lead your mind to automatically drive you home.

That's why visual aids—graphs, pie charts, four-color charts—get through to people what numbers and sentences do not. So a refrain some philosopher friends used to sing, "I've Got Dem Mind–Body Duality Blues," isn't really so. This hints that our fears of disembodied intelligences are a bit off the mark. Mass media science fiction sometimes uses distant, emotionless computers as villains, as in *2001*'s HAL and *I, Robot*'s vast computational complex. These are a convenient dramatic shorthand—nobody questions the motivations of machines—and express deep uneasiness. But a really effective intelligence will have—and presumably enjoy—a constellation of body-sensations, just as we do. That's a *long* way off from our present understanding of machine capabilities.

And what does it mean to be humanlike, anyway? One of our mass media preoccupations is a rather sentimental reverence for emotions, as if they were more basic than mere dry thinking.

To be in the grip of a strong emotion is to "lose reason," to feel totally focused on a narrow range of events. When a dire situation arises, emotions take over with dizzying speed. Will I (the Self) escape from this fire? Will I find food? Stay warm? Make a touchdown?

Do we have emotions—or do emotions have us? That is a deeper question than you may think, because proponents of artificial intelligence (AI) have conspicuously neglected the study of emotions. Most seem to feel the subject is just too complicated. Compared with, say, reasoning your way through a theorem in geometry, or picking up a glass of water without spilling it, emotions seem messy, dangerous, hard. After all, our arts and culture mostly concentrate on them.

Could it be that emotions are in fact simple? That they are mysterious to us because we have a blinding flaw in our method, a defect in attitude that makes the way to understanding hard for us to see?

Some ponderers of our Selves are beginning to suspect the second answer is nearer to the truth, including Marvin Minsky of MIT, who remarks, "We think our feelings are elemental, basic, but they probably aren't. They may be learned." He points out that emotions come from truly intricate machinery, and that the power of emotions over us doesn't necessarily mean the underlying causes started out as complex. Instead, they may begin at birth with simple but strong drivers that get elaborated by us as we grow up.

Minsky feels that it's the power in emotions that deceives us. Maybe they are "wired up" to sources of forceful inner control, and we're dazzled by that?

Suppose we begin life with some simple building blocks in our brains. These nearly separate drivers—Minsky calls them "agents"—have virtually autonomous control of a specialized range of responses. Evolution would shape them to have clear-cut goals, centered on survival.

If the baby runs low on sugar, the "hunger" agent steps forward and seizes control of the conscious and subconscious levels, piloting our attention toward the pressing problem of finding food. Maybe that just means opening your mouth so your mother knows to feed you. Later, it may mean conveying this information with facial gimmicks. When you're still older, the hunger agent learns that it can charm some food out of mother or father with a smile, or a gesture, or (much later) a convincing argument about why you really deserve that extra cookie.

The same learning operates for your "cold" agent, which learns to move into the sun, or cover itself with insulators, or snuggle up to someone. This last maneuver proves to have side benefits, bridging the chasm between the baby and other people, and eventually linking up with the "reproduction" agent.

In this sense the emotions are agents narrowly focused on solving problems—they're rational. What makes them seem irrational to us? Because they're often in conflict with forces we've learned about in the world, or even with other agents within us.

We ride herd on our emotions as we age—in fact, that's the usual definition of maturity. This means that the agents get more savvy about reaching their goals. Don't cry for food—charm someone who has some. Babies who learn to smile early get a reward-stimulus of gushing attention, food, the works—which means that a few generations downstream, there'll be more babies with that skill, and fewer who wear a permanent frown. Later, the baby will learn to earn money for food, but by that time its many agents are a well-managed crowd, intent on their little jobs.

You become intelligent, in this view, when your agents collaborate to form an effective team. Mild likes and preferences bring into play cooperation with agents who may have something else in mind. Spending money on a date leaves your "hunger" agent without the cash to indulge in a filet mignon feast.

Sometimes we're aware of our agents. People strike bargains with themselves. "I'll work another thirty minutes before eating that doughnut," or, "If I don't earn more, maybe she won't love me." (Agents don't have to be rational, in the sense that their strategies fit the external world perfectly, or make sense to other people when expressed. They won't be well socialized, either. They're basically just smart appetites, after all, not philosopher kings.) Knowing you're confused is actually a high-level activity: evaluating your own states, "feeling" your agents battling with each other.

Agents start out as crude directives. Neural connections in babies elaborate themselves, starting from simple patterns of wiring that only grossly resemble the usual adult pattern. Although we're born with almost all the neurons we'll ever have, a baby's brain is only one fourth the mass of an adult's. The brain swells because neurons grow, making many more connections. Stimulus makes this happen. Babies left lying in cribs develop slowly, while ones who get a lot of attention develop skills much earlier.

So the brain learns how to wire itself. The higher up the evolutionary pyramid a vertebrate is, the longer this construction takes. It's easy to understand why evolution would favor such on-the-job training: it's genetically conservative.

If our genes tried to exactly specify each neural connection—hardwiring us with molecular markers—there would have to be an instruction manual in our DNA. It isn't there—the DNA doesn't have enough storage space to carry a Library-worth of circuit diagrams. Using activity-dependent remodeling is far more economical with DNA space. It also allows us, and other higher animals, to fit ourselves to the environment better.

The simple agents we start out with have only basic instructions, then. That could be why babies jump so suddenly from blissful calm to irritable hunger or blistering anger. Older children don't skip so quickly through the menu of moods, and their faces show a more complex blend of responses to the world as they age—what we call "character" in a face.

Those beginning agents could then develop semi-independently, storing up private memories of past incidents, learning from the world *as they perceive it*. They also hear about other agents that can have conflicting needs. Appetites jostle for advantage, slowly learning to use each other productively.

There is considerable evidence that different drives emerge from distinct parts of the brain—the site where an agent "lives," perhaps. Neurologists have classified about a hundred different processing regions in our brains, each a neural subnet.

Emotions are mysterious to us because of their great power over our sense of reason, but there can be plenty of other kinds of agents, all tutored by experience. Probably it takes several agents to carefully cut up your dinner with knife and fork while you're arguing politics, for example—and another agent is simultaneously keeping a vaguely wary eye out for any disturbance coming in from afar, like a fistfight at the next table, or an earthquake.

Each emotion-agent can feel to us like a different personality acting in our minds. Indeed, to be seized by anger, say, is to be turned into a simple machine, with clear goals ("Hit him!")—though other agents can wrestle for control, especially after the central desire is satisfied. ("Now that he's down, let's run for it!")

Emotions are powerful because they tap into deep resources. They can command the stage of consciousness, because they were once crucial to survival. But whom do they wrest control from? The Self—but who is that? This is what Minsky calls the Fallacy of the Single Self.

We feel that somewhere, deep down in the marshy Mind, there's somebody who is Us. We're the stage upon which our internal dramas act. Suppose, instead, the Self is an intangible thing in the same sense that Beethoven's Fifth is—a symphony of many instruments, expressing themselves in concert. Emotions are hard to contain in the same way that a mob can get out of control—there's no dictator.

Freud's "superego" was the Old Man who made you do the socially okay right thing, but the Old Man obviously doesn't work well all the time—or else soap operas wouldn't exist. In this view, what makes the Human Condition—a cliché that nonetheless means much to us—is that we vaguely sense the constant warring agents that make our Selves. To be in complete control of yourself, then, is to reduce yourself, to let one agent rule.

We don't hold absolute sway over the agents, because we *are* them. And they don't communicate well. We normally have only a partial idea of what is going on inside ourselves. We don't consciously know how a sentence is going to end when we start it— that job is done by some club of agents who operate beneath our conscious perceptions.

It's easy to see how such an unconscious, collaborating club could evolve—it's much more effective to be able to run fast, holding a spear ready to throw, and also, without a lot of bother, call out to your fellow hunters, telling them to back you up.

Of course, we like to think that we know ourselves. We have our own internal worlds, knowing the ache of a stubbed toe with a poignancy that surpasses our sorrow at, say, the deaths of a thousand people on the other side of the world. But of course we notice the ache only when it crosses a threshold of pain, and are glad when it eases off, falling below that threshold. We often don't really know we're tired until we "discover" that we're yawning—some agent has seized control of our breathing, down in the medulla, and made it deliver more oxygen.

So maybe the distinction between reason and emotion, beloved to us all, is the wrong set of polar opposites. If agents evolve as we mature, maybe "simple" versus "complex" is a better way to think about the non-single Self. Reason is more nearly associated with a well-governed society of mind—an orderly legislature. Emotion occurs when a faction seizes the parliament and forces through actions that satisfy short-term interests of a minority—in babies, a single agent.

A large body of experiment and careful, step-by-step theorizing—unlike the conceptual leaps in Minsky's *Society of Mind* overview—are leading us rapidly toward a new understanding of ourselves. No longer do neurologists separate seeing cleanly from understanding, or even visual comprehension from consciousness. Our potent combination of moment-to-moment awareness, plus quick retrieval of memories, makes up the working "blackboard of the mind" that makes us plan and act better than any other species.

There's a dark underside to all this enlightened progress, though. A biologist friend recently remarked that when he shows visitors to his lab a worm's tiny knot of nervous tissue, and describes how we can now follow all the command routes to wormy thinking, nobody is much concerned.

So he shows them a common housefly. Already he has found the neural wiring that makes the fly weave its erratic course through the air, a feat of piloting they "learn" in moments. He can lay out the specifications of this autopilot on a circuit diagram. This doesn't disturb his guests, either.

But show them a section of a human brain, and point out that a small wedge of it looks pretty much like the common housefly's hardwiring—and you've got trouble.

Determinism. Who likes to think that he or she is simply an elaborately detailed robot, following instructions that genes and happenstance have laid down?

Yet much research strongly suggests that the brain is not the tool of the Mind, or a house wherein Mind lives, but the master of Mind. Damage or deprive the brain and you alter Mind. Introduce a current into the brain at the right place, and sensations or even memories come flooding into consciousness. A simple materialistic reading of the data implies that we are biological machines, because that is the operating paradigm, the reductionist instinct, of science.

No higher mental functions are in principle immune to our modern analytic, atomizing frenzy. A recent book, *The Biology of Religion* by Vernon Reynolds and Ralph Tanner, holds that "Religions . . . act as culturally phrased biological messages . . . a kind of 'parental investment handbook.'" Similarly, the anthropologist Lionel Tiger (wonderful name), holds that natural selection responded to "an age-old problem: what to do with a cerebral cortex that has the capacity to create immobilizing scenarios of disaster and to dwell fruitlessly on the utter meaninglessness of it all." The solution: "wiring into our brains a moderate propensity to embrace sunny scenarios even when they are not supported by the facts."

He has some evidence, too. "People remember 'up' words— 'happy,' 'attractive,' 'bright'—in preference to downers. They elect fuzzily optimistic politicians over painfully candid ones. . . . One thing likely to keep the endorphins flowing is for everyone to get together and agree on the story that someday everything will get better." Religion he includes among these ideas.

Tiger's ferocious conclusions represent the reductionist wing of current sociobiological theory. As we accept more and more the computer-like analogy for our Selves, perhaps we will come to see consciousness as a necessary buffer between us and the world, no more.

This kind of shift in thinking we already see in pleas that people, especially the poor, are victims of society or other influences, not responsible for their own failings. Soulless machines can be neither heroes nor villains. Of course, neither can they hold inalienable "rights," some will feel, or take credit for their accomplishments—if all is determined by genes and accident.

Well-meaning scientists have shown already that prenatal stresses and "bad" genes cause later health and mental disorders. Our beleaguered psyches will generalize from the accretion of such self-diminishing truths, and no doubt find it all quite depressing.

But such conclusions are both hasty and oversimplified. Abandoning the model on the Single Self, as Minsky advocates, can free us. If our agents do indeed learn and grow, that implies a complex relation between mental events and electrical spurts along axons and dendrites. We are *self-programming*—so a simple deterministic picture doesn't work. The Single Self, which sat like an unmoved mover at the center of our minds, dissolves, replaced by Mind as an emergent property—one that cannot be accounted for solely by taking the component parts one at a time.

In our bodies, the heart beats because its pacemaker responds to the ebb and flow of certain ions. But the pace cannot be fathomed without referring to the effect of the pacemaker on the flux, too. Such interlocking systems may yield consciousness as an *emergent* property, which cannot even in principle be predicted in advance— and therefore cannot be known to be deterministic. It is far too early to resurrect the "free will" arguments of the last century—and some will find that liberating.

"Determinism" usually means *pre*determining the outcome of events. A Self of many agents cannot be so predicted because—as chaos theory has taught us—in such complex interactions, detailed outcomes cannot be found even with completely specified initial conditions. The best we can do is predict the kinds of outcomes a Self might reach—say, deciding whether to have chocolate or vanilla ice cream, given the Self's past. How much would we need to know to make even this simple prediction possible?

We won't have answers to such issues for quite a while. To take ideas about Self and consciousness further than the maybe-this, maybe-that stage, we need to work out simple models. One would like to see computer programs explore simple situations, following the kinds of rules that "artificial life" research uses.

Take three basic drives, say, and rules for their expression. Let agents for them learn and remember, but communicate between each other only poorly. (One of the crucial ideas here is that you can't let basic agents override each other too easily. Pursuing, say, sex and forgetting about hunger could be fatal.) See how well they work, grow, and "mature." Whether this approach will reveal more than the biases of the programmers is always a gamble. But it's a start.

One thing is fairly clear, though—consciousness is natural, has deep biological roots. Just as animal bodies emerged from evolution's persistent winnowing, in which chance variation is pruned by natural selection, so animal minds must have some ancient origin in the intricate mechanisms of differential survival, working on the available materials.

By "animal minds" one may mean simply that anybody watching a dog or a chimp figure out a problem will recognize signs of an intellect confronting an external world, manipulating it, and storing the fresh information. That's intelligence. Chimps, in fact, can rearrange sentences with the skill of a two-year-old human—but curiously, they don't have the subject–predicate sentence-forming talents we manifest naturally. And they apparently never rise above that two-year-old level. Evolution never rewarded chimps for moving in that direction in their language-shaping tool kit.

Consider a larger question: How did mentality arise from cell tissue? Consciousness, defined as some model of the outer world, is prevalent in the biological world. Above the simplest organisms, many animals seem to have some smattering of it. One readily can think of two general reasons that might explain this, neither compelling.

Maybe consciousness has some huge, unique utility. On the other hand, could it be written into the very nature of matter itself, and can't help emerging when particles coagulate in any of a wide variety of ways?

The first possibility is appealing, but it seems easy to imagine even complex organisms making their way, mating and foraging, without having to be guided by sentience. Nobody home but us expert systems here, folks.

A counterargument to this observation is that when we look at our abundant natural world, we don't see complex, robotlike species. The insects are simple robo-species, and vastly successful at it, too. Why haven't they inspired higher-level organizations? Why don't we see larger, more sophisticated animals who don't seem to have self-awareness? Maybe the talent for making representations of the outer world, not merely taking in signals, is so powerful a tool that nature has found no other way to perform really refined tasks.

So maybe consciousness emerges as organization increases in the brain, period. But suppose consciousness is simply a by-product?

As we climb up the evolutionary ladder, and the ratio of brain mass to body mass rises, many traits emerge that also have great use—better coordination, vision, weight, and the like. Could we be missing the *really* significant facets of evolution? Just for the sake of argument, perhaps *vision* is the key, and consciousness is just a great way to make better use of eyes. A big fraction of our brains is involved in seeing, after all.

Indeed, one of evolution's great puzzles is why sentience seems to be the preferred method for handling adaptivity to the ever-changing environment. Why not process information without any inner feeling at all? Why does consciousness exist?

The second possibility, that sentience emerges inevitably from the substrate of matter, essentially relegates the issue to sciences other than biology. This implies that there need not be some innate processes in evolution that work toward consciousness. That would be good news for the SETI folk, who tirelessly Search for Extra-Terrestrial Intelligence, since it implies that Mind will arise in myriad physical milieus. But as an explanation, biologists find it unsatisfying, since it leaves them out of the action, relying instead on vague notions of self-organization arising from some deeper organization. To say the least, this issue is wide open.

There is an even more unsettling possibility—that we cannot fathom consciousness because of our own limitations. Our human power of comprehension is itself a natural, biologically derived ability, constrained by facets of our origins.

Our minds have limits. There is no automatic reason why our capacity to understand nature extends to all things that puzzle us. As philosophers have long known, the ease of asking questions does not imply an answer. It is a real possibility that our biology has not equipped us to grasp our own consciousness, because the job entails whole categories of thought unavailable to us.

After all, we probably weren't "designed" to know ourselves, in the sense of seeing how we work mentally. You cannot watch yourself have ideas—they just happen to you, welling up from that subconscious that does so much of the real work in our lives. The capability to watch your mental machinery turn its cogs and gears would have little utility in a hunter-gatherer society; it is not obvious that it would have any real use even now, except perhaps to professors who could settle some intriguing research points. We speculate on such issues because we are playing very different games now.

This line of reasoning is forever uncheckable, of course—it speaks of boundaries we cannot see. Closely allied is the idea that genuinely alien minds would act in ways we could not recognize. The radio-listening SETI community is gambling that this is not so, but one wonders. . . .

Will truly advanced minds appear to us as natural, though perhaps vast, phenomena? Ants do not ponder the beings whose shoes crush them, after all. And perhaps that's for the best.

These issues will affect our own augmentation. Mathematician Vernor Vinge envisions a time when some portion of humanity will progress suddenly, using our computer and enhanced biological technologies. By gaining powers that exponentially get more effective, an advanced community could move through this "Singularity"—beyond which we mere mortals cannot see what happens next. Think of an ant crawling across an opened encyclopedia, wondering at the mysterious squiggles on the page.

Will we augment ourselves to that state? How would we know? If we see other people doing grand, inexplicable things that appear beyond human powers—yes, that would indeed be a clue. But simple societies have had that experience already, when aircraft roared through their skies and apparent demigods emerged. It did not take long for them to see that these interlopers were mere men, after all. Probably those left behind by the Singularity will see it as a challenge, too.

Still, science fiction has spawned many visions of transcendence, often with theological trappings. Angels, after all, are images of beings of divine origin, with vast powers. Perhaps our future could give us beings who would seem angelic to us today. But let us remember that Lucifer was an angel, too.

Man Plus & Plus & Plus & . . .

Where does an ever-better, ever-ready body lead? Can we live for centuries? Is death really inevitable? Who decides?

First, a few distinctions.

"Android" means "manlike," but that includes both humanlike forms made of organic substances and machines that look like us. Ultimately, it may be hard to tell which is which, and the term is used casually, since either form is still in the future. Sometimes, robots with human shapes are termed androids, or 'droids, even if they have few complex human characteristics. A moving mannequin is the simplest form of android.

"Robot" has similar wide-ranging uses. The originator of the term, Karel Čapek, coined it for organic humans grown in vats as mass-produced slaves (something like fictional clones). In his 1920 play, *R.U.R.* (for Rossum's Universal Robots), the robots are made so nearly like us that they have souls, and eventually overwhelm humanity. In Czech, Čapek's language, the word means something like "serf laborer," and this play introduced the basic theme and anxiety surrounding the subject: that intelligent beings made to help us will turn on us, proving in the long run better than mere humans.

This portrayal began a long tradition of authors taking the side of artificial forms. Throughout most twentieth-century science fiction, the discussion assumed that robots, not cyborgs, would be the first of the new, improved beings. Usually robots are seen as an oppressed class, in narratives about racial or ethnic issues only thinly veiled. Robert Silverberg in his 1971 short story "Good News from the Vatican" saw a robot chosen pope, apparently the highest robotic ambition. (And who better to take one's confession?) Many fictional and film robots gradually acquire human prostheses, blurring the boundaries still further; this was the point of Isaac Asimov's short story (later a film with Robin Williams) "The Bicentennial Man." That story took a robot's ambition to become human as a natural outcome, implicitly assuming that humans are the pinnacle. Even a machine that was stronger, smarter, and far longer-lived than any human it had met would or should strive to be like humans. Such self-congratulation is common in science fiction.

Though Asimov wrote "The Bicentennial Man," he also depicted in his collection of short stories *I, Robot* some robots who doubted that mere humans could possibly have invented them. They argued that they were far better in every way than humans, who were probably lying about where robots came from. So some unseen, divine being must have made them—that is, God.

"Cyborg" is a shortening of "cybernetic organism" and "cybernetics" is an older term for the abstract study of intelligence, particularly of machines. In practice it means hybrids of man and machine. (One relevant example is the Borg of *Star Trek,* whose slow-witted collective social logic seems to vary with each episode.)

Cyborgs have been with us a long time—Peg-Leg Pete the Pirate with his wooden leg is one, and technically, so are those of us with eyeglasses. The essence of a cyborg, though, is the feedback between human and machine, from servo-mechanisms to turn a wrist (now under development) to fully automated limbs. These augmentations are far more personal and integrated than eyeglasses.

The term "cyborg" was first used by Manfred Clynes and Nathan S. Kline in 1960 to describe a future breed of astronauts, engineered to withstand the rigors of long-distance space travel. With lungs replaced by nuclear fuel cells and their hearts controlled by injections of amphetamines, they would be rugged and space-adapted. This usage has paled beside the prospect of people augmented for high performance on Earth.

Three classes of cyborgs emerged in film and fiction, though they are only beginning to be realized in hardware. First of these are the adaptations humans accept to work in different environments (usually alien worlds). Less radical are people rebuilt to perform specific tasks in tough places (the hostile Detroit of the film *RoboCop*). Either way, they can't go back to being human again, and their anomie at their separation from humanity is often a major plot element.

The most radical idea is that human minds would be installed in far grander mechanisms. In Samuel Delany's novel *Nova* people plug themselves into automated factories which they oversee and direct. A step up is Anne McCaffrey's *The Ship Who Sang*, which depicts an entire spaceship run by a woman's brain, fully integrated so that it *is* the ship. In these cases, rather than suffer anomie and isolation, the characters feel powerful, grand, and exalted—a far more positive vision.

"Bionic" is a contraction of "biological electronics" and includes any electrical addition to the already functioning electrical aspects of the body. Coined in 1960, it more generally describes making artificial systems that resemble living ones. This can mean simple appearances, as in the eighteenth century, when automata were made for amusement, including a duck that swam, flapped its wings, and quacked.

Using natural models is a good beginning. After all, plants store sunlight in chemical form with nearly 100 percent efficiency; our nervous systems transmit information better than telephone systems; muscles are highly efficient motors; our brains exceed the capacity of present computers. The U.S. Army has experimented with mechanical legs copied from insects, to aid both carrier machines and soldiers in movement through rough terrain and swamps. The machines worked reasonably well, and some soldiers were able to move better than with their unaugmented legs.

However, simple imitation of nature ignores the difficulty of translation. Airplanes do not fly like birds, and could not. (Though helicopters do use the basic mechanics of maple-tree fruit, which spins as it falls, and can remain aloft to travel long distances in a fair wind. This was knowingly copied by Leonardo da Vinci in his sketches of helicopters.) The better course is not to copy in detail, but to fathom the principles underlying what works in nature, and why it has been selected by evolution.

A bionic woman would not necessarily be stronger or more agile than a normal woman, but universally the term has come to imply amplified abilities. The term neatly gets around our wariness about strictly biological experimentation with ourselves. We shy away from modifying the genomes of later generations, but not from working upon ourselves.

Typically people find it more acceptable to take into their bodies a fresh metal shoulder socket than a synthetic organ. Electronically enhancing ourselves, perhaps even mentally, seems more machine-like and hence less threatening. The Bionic Woman, centerpiece of a TV series (1976 to 1978), had a bionic ear able to eavesdrop from miles away. The Six Million Dollar Man, her erstwhile boyfriend, had a bionic eye with similar range. In all cases, bionic means improvements upon normal abilities, but nothing truly new, fundamental, or disturbing. (No one ever asked what bionic genitalia might mean, for example.) Television is usually not disturbing, so this conservatism is no surprise.

In common use, all these terms blend together. Generally, mechanical systems that do not resemble humans at all are simply regarded as machines, and not assumed to have human responses. The computer, HAL, in Stanley Kubrick's *2001: A Space Odyssey* controls the entire ship and its mechanical systems, using them to kill nearly all the crew. But HAL remains a machine, and we do not regard its parts as robots.

Distinctions about humanoid forms will be much tougher, because we automatically begin assigning human nonverbal signals to shapes that suggest human ones. People read facial expressions into clouds, Rorschach blots, and random bits that have the signature mouth and eyes positioned roughly right. A cyborg with obvious machine parts will probably find many who will listen when it demands human treatment.

Given some level of acceptance, how might cyborgs affect us, our society, and their own evolving status?

Immortal Cyborgs?

The oldest person whose age is reliably known was Jeanne Louise Calment of Arles, France, who lived from 1875 to 1997, achieving 122 years 5 months. As a girl she had sold pencils to the young, unknown Vincent van Gogh. Aside from such exceptional cases, the lot of humanity is shown by the mortality curve we give here. As it shows, there are two major stories taught by history about improving longevity.

First, the dramatic improvement of modern times has come mostly from the better survival rates of children. In a state of nature, children fall prey to cold, disease, and accident at high rates. Better sanitation, medicine, and nutrition have made their gains throughout the twentieth century. This progress explains why the

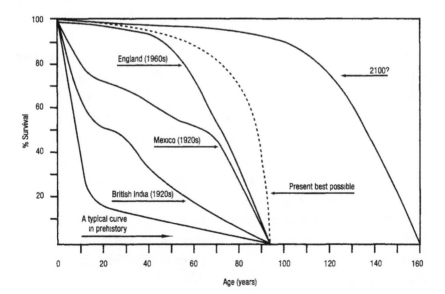

curve for Mexico in the 1920s differs greatly from British India at the same era: India had yet to enjoy the improvements slowly diffusing from the advanced nations, while Mexico was already benefiting. Note that the difference between England of the 1960s and Mexico of the 1920s came mostly from better child mortality rates.

Second, the elderly death rate has shown some improvement, but not a lot. There is still a fairly solid "wall" around age eighty, and beyond it, the population declines roughly exponentially. One might term this the "fragility wall," where people become prey to any passing microbe or severe accident. Their resistance and resilience have eroded until they are easy marks.

We all take extending longevity as a good to be sought. Some will carp about increased costs to Social Security, or population growth, but getting more out of a single life also promises huge gains, many of them economic, as people work to greater age. Given the search for more years, it is surprising that there is so little research into the deeper questions of how to push back the age frontier. Is there a basic limit? And is it the same for us all?

Notably, since 1900—when the death rates of the sexes were just about the same—women have consistently gained extra years of longevity over men, until now they live about 10 percent longer in advanced societies. Few take note of this remarkable inequality, which is still increasing after a century. Females consume more than two-thirds of health care budgets, and are consistently heavier users of health services throughout their lives. Upgrading male longevity to the level of females' in the advanced societies would improve the average human survival more than, for example, completely eliminating cancer. This strongly suggests that social forces have a great deal to do with improving our expected life spans, beyond the reach of technologies alone.

The other major factor affecting longevity is prosperity. Rich societies fare better through their medical technology and through generally better diet and habits. Today in prosperous nations low-fat diets, little smoking, and conscientious exercise can extend longevity. Most of these improvements come from better education, not from technology.

Will augmented people have longer lives? Even obvious mechanical aids like better legs and hips could prevent the often dev-
astating falls among the elderly. Certainly radical technologies like
nanotech would profoundly affect longevity, allowing replacement of
cellular materials and direct, pointed interventions in major causes
of death today, diseases like cancer and arterial blockage.

The figure's curve labeled "present best possible" is a guess at
what might be achieved by present technologies on all fronts.
Shoring up the elderly could plausibly lead to the dashed curve
within about fifty years.

After all, in advanced nations the average longevity has increased 50 percent in the last century. A similar improvement
would take us to a curve that terminates somewhere between 100
and 110—a cotton-topped future. But technology changes, and the
advances from augmentation plausibly can keep making inroads on
the many causes of our mortality.

The figure's curve labeled *2100?* is of course a pure guess, building on the successes of the past century. It assumes the "fragility
wall" around age eighty has been thoroughly broken down, with another 50 percent increase beyond the "present best possible" curve.
This line is not a serious prediction, but rather a suggestion of how
much augmentation could change us. Certainly replacement organs
(transplants or outright mechanical organs) can prevent deaths.

As well, such extensions begin to change our views of the human condition. We already see that young people are delaying education, marriage, and other social goals beyond those typical of
people in the first half of the twentieth century. This postponement
may come from their sense that they have plenty of time, since they
see their parents leading vigorous lives into their seventies and even
eighties, a phenomenon nearly unknown only a century ago. Such
subtle changes go unremarked because they are intuitive.

How far can this go? We have no true idea of an upper limit on
life span. If we eliminated all aging, so that we faced no fragility
wall, eliminated diseases, and could avoid all causes of death except accident (including suicide), how long could we live? Most people, when asked, guess at ages like 120, or 150. The answer, gathered

from studying the causes of death in actuarial rate tables, is astonishing: close to 1,500 years! This seems more plausible when one
reflects upon how many friends die of accident. Typically, one
knows only a few who die in accidents before age fifty, from a total
of, say, a thousand friends. This number translates to a death rate
from accident of about 1 in 1,000 per year, or an average expected
life span of about one thousand years.

With only a century or less of life, humans have developed
many social forms to deal with that span, and nearly none that look
beyond it. Take just a small step into that immensity: Imagine living
to 150. How would you plan a career? Could you keep interested if
the job (like most) had a fair level of routine? And what about marriage? Some argue that the divorce rate is high these days because
people know they face a far longer span together than they did a
century ago. Perhaps marriage itself can be redefined to set term
limits, an idea that was called "contract marriage" in the 1940s and
never caught on. What of children, as well? As we live longer, the
population growth problem worsens if we keep reproducing. Our
lives will be less family-centered because our children will be
adults, with their own lives, for a larger fraction of our spans. New
social problems may be the most profound implication of the path
the cyborg will lead us to follow.

Radical Cyborgs?

Much thinking about both androids and cyborgs has projected
upon them a political and social agenda—that of revolutionaries.
Donna Haraway's essay "A Cyborg Manifesto: Science, Technology,
and Socialist-Feminism in the Late Twentieth Century" (1985) held
that the prospect of augmented bodies could "dismantle the binary"
in human culture—that is, the assigned sexual and gender roles.
She felt that the new body changes and information technology
could erase both gender and racial boundaries and the "structures
of oppression" that have historically gone along with them.

Certainly the prospect of making labor less dependent upon
motor muscles, with computing skills emerging as more important,

seems to promise a more nearly equal footing for women in the workplace. Perhaps the trend would go beyond that, to cyborg bodies with mixed male and female characteristics—buff features, sinuous curves, and sexually ambiguous design features, available to all at a reasonable price.

Experience has not been kind to this idea. Where biology has extended our reach, women have flocked to the technologies that extended their ability to bear children, and thereafter have lived remorselessly conventional lives. Even homosexuals have used reproductive technologies to create their own nuclear families, not new constructions, "alternative societies," and homes with experimental modes of living.

Some feminists have seen the constraints of the body as aiding forms of "ideological containment," meaning that biology equals destiny, in the social sphere especially. But consider popular entertainment, often the surest way to discover what the public will accept. Cartoon superheroes are now equal-opportunity bruisers. Women superheroes are common now, superfemale in appearance just as the men are supermale. But they are united in holding that aggression, muscle strength, speed, the killer instinct, and solution through destruction are always the right answers—hardly a signature of femininity.

Interestingly, the popular Japanese cartoon film *Ghost in the Shell* treats cyborgs as "shells" that can have mixed sexual characteristics, appearing female in some views and male in others. The subtext is that to become a "truly living" being the cyborg must display sexual traits, but can choose from either as the situation allows. Of course, to be both is, in the minds of many, either to be repulsive or to be neither.

Humans are nowhere more sensitive than in matters of reproduction, and therefore sexuality. It seems unlikely that cyborgs or androids will carry forward a new agenda, but rather that augmentations will shore up the desire of people to fit into the norm, to not stand out more than they must. Few individuals want to appear, either physically or socially, as a "freak."

It is one thing to stride down the street looking outlandish, perhaps with a half-ceramic head sporting a wearable computer, read by pixel-augmented lavender eyes . . . and quite another never to find a mate, because no one of the right sexual polarity has chosen your special style statement.

Cyborg Consciousness

What will happen in the psyche of a cyborg? Will changes and augmentations grow to the point where the human is swamped by the machine? Would a truly enhanced wearer begin to resent the puny humans who service his bionics?

We take our biological body for granted, for the most part unaware of how it shapes our thoughts. But we are united with the rest of humanity by the biological limits of our fleshy existence: birth, development, reproduction, aging, and death. We have two parents and can produce offspring only with another person. We have to eat other organisms, thus we have to tend them. We have to breathe a certain mixture of gases, live within the same narrow range of temperatures. We need to sleep. We die at less than one hundred years of age. Our physical and mental capabilities are roughly the same as the next person's, or at least within a familiar range. Certain substances will poison us, and so on. This commonality produces an extremely deep kinship with other human beings, and to a lesser degree, with other animals.

Dr. Anne Foerst, a theologian, works at the Artificial Intelligence Laboratory at MIT with scientists building two robots with humanoid features—Cog, a humanoid torso with arms and a head, and Kismet, the MIT face robot. You might say she is a robotheologian. She agrees that our particularly human intelligence is strongly molded by our human body experience.

> In order for a machine to really be intelligent, it has to be embodied. We say intelligence cannot be abstracted from the body. We feel that the body—the way it moves, grows, digests food, gets older, all have an influence on how a person thinks. That's why we've built Cog and Kismet to have humanoid features.
>
> Cog moves and experiences the world the way someone who can

walk upright might. He experiences balance problems, friction problems, weight, gravity, all the stuff that we do, so that he can have a
body feeling that is similar to ours.

Cyborgs have different realities from ours. What if we were a
brain encased in a metal body, unable to feel the touch of others or
reproduce, but with a vastly extended life span? Would we lose sympathy for the mass of all-flesh humans? And how much machine
could we be without losing our humanity?

Hollywood has taken on the fusion of man and machine often,
in such films as *RoboCop, The Terminator,* and the TV series *The
Six Million Dollar Man* and *The Bionic Woman.*

In *RoboCop,* a fatally wounded police officer, Murphy, is resurrected as a practically invulnerable cyborg. His memory erased, his
human ethics are replaced by four directives to govern his behavior
(an idea borrowed from Isaac Asimov's Three Laws of Robotics). At
first he functions well as the invincible cop-machine envisioned by
his corporate creators, but when his former partner recognizes him,
he starts on a quest to regain his humanity.

The quest for humanity is explored also in the movie *The Fly.*
In the original film, the hero is a scientist whose body is part fly after an experiment gone awry. The fly part becomes steadily stronger,
and his last human act is to kill himself before he harms his family.

The human/machine dichotomy theme is not explored in the
earlier TV series. *The Six Million Dollar Man* and *The Bionic
Woman* are simply episodes of whiz-bang feats of strength, speed,
and visual or acoustic acuity by the bionically enhanced humans.

In *Star Trek: The Next Generation,* humans encounter a collective intelligence known as the Borg functioning as a single mind
with many mobile units. The analogy is roughly that of an ant
colony acting as a single organism (although the concepts behind
this TV creation can be altered by scriptwriter whim and change
over time). The Borg scour the galaxy for organic bodies to add to
the consciousness. To achieve mind control over humans, an electronic device is plugged into one side of the head, usurping the
function of one eye, and converting an individual human into a

cyborg. Because the main story line concerns the (horrific to humans) collective mind theme, there is no discussion of the implications of the man–machine fusion as such.

At one end of this continuum is the wholly artificial android of Asimov's "The Bicentennial Man." He longs to become human, undergoes a series of upgrades exchanging flesh for metal, and finally achieves mortality. This idea is the basis for a *Star Trek: The Next Generation* character, the android named Data.

These stories share the comforting theme that as machines become more sophisticated, they will inevitably try to become more humanlike. In *RoboCop,* we sympathize with mostly artificial Murphy when he takes off his helmet to reveal a face of flesh. He then goes on to defeat the entirely artificial robot that symbolizes his mechanical parts, thus affirming his kinship with the rest of humanity.

But that may not be how the future plays out.

A fresh angle on cyborgs appeared in Damon Knight's classic 1968 short story, "Masks." The point-of-view character is a cyborg in a new body, the focus of a whole research group, who is displaying emotional problems. He refuses to reveal the cause of his troubles to those working on the cyborg project, but in a scene near the end, we enter his thoughts:

> No more adrenal glands to pump adrenaline into his blood, so he could not feel fright or rage. They had released him from all that— love, hate, the whole sloppy mess—but they had forgotten there was still one emotion he could feel.
>
> Sinescu, with the black bristles of his beard sprouting through his oily skin. A whitehead ripe in the crease beside his nostril . . .
>
> Babcock, with his broad pink nose shining with grease, crusts of white matter in the corners of his eyes. Food mortar between his teeth.
>
> Sam's wife, with raspberry-colored paste on her mouth. Face smeared with tears, a bright bubble in one nostril. And the damned dog, shiny nose, wet eyes.

Note the repeated use of the nose, probably not our best feature, as inherently ugly. Later, in the story's last line, we feel how this cyborg wants to get away from all things organic and fleshy, to become wholly metallic, on the moon: "And he was there, and it was not far enough, not yet, for the Earth hung overhead like a rotten fruit, blue with mold, crawling, wrinkling, purulent and alive."

Will cyborgs then do the opposite of their portrayal in so many stories, and try to be nonhuman? We would term that a "disorder," but it could seem utterly natural to one who felt the cool, clean beauties of his new state; a fresh, new being of unknown impulses.

Of course, matters need not go so far to have major social effects. Cyborgs may feel themselves to be sufficiently different from flesh-and-blood humans that they will draw away, perhaps repelled, and form their own interest group. Mostly metal cyborgs may have different concerns about their environment than humans. Cyborg politics will emerge.

This option is not so far-fetched as it might seem. Groups representing persons with disabilities have become very active in recent years, in the United States and Europe. As a result, wheelchair access to everything from public bathrooms to castles and beaches has become an important marketing consideration.

In the future, we can look forward to cyborg political action groups.

Humans are group animals. We live together, not alone, and we are powerfully drawn to others with similar attributes. We find the company of others comforting, an emotional necessity. It's also part of our survival strategy, and has been in our genetic makeup for millions of years. It is no coincidence that individuals who commit senseless acts of violence, like shooting at strangers, are usually loners.

Brain transfers into titanium bodies are admittedly far off, but what if an individual possessed significant mechanical or electronic parts? Neurological implants, perhaps, to correct stroke damage or catastrophic spinal cord injury. We are now entering the realm of the true cyborg, because in order to function, their implants would have to be permanent.

Undeniably, the wearer would feel instant sympathy with others bearing similar implants—sharing the color of a sunset as seen through cameras rather than eyes, for example. Wary of being considered as freaks by the rest of humanity, they would draw into their own group.

Already Deaf activists celebrate their deafness, calling it part of normal human diversity, and not a handicap. They communicate via ASL (American Sign Language), which they describe as a true language, complete with grammar, and they see the use of modifications like cochlear implants as a threat to "Deaf culture." In the fall of 2000, the documentary film *Sound and Fury* explored these issues on the screen, and a character in the TV series *ER* opted not to have his young son receive cochlear implants.

How many implants would it take before a person stopped being human and became a cyborg in the eyes of the world?

Maybe fewer than we imagine.

Professor Steve Mann from University of Toronto, who has worn an experimental personal computer (not even an implant) since the 1970s, noticed that people avoided social contact with him unless the device was small enough to be overlooked. His wearable computer, while allowing him access to everybody on the Internet, cut him off from the people next to him. He looked different. Even cell phones have this effect.

The old adage, "Birds of a feather flock together," is also true about us. Will we hear calls for bans of cyborgs in restaurants? ("People are trying to eat here!") The "robot revolt" in Karel Čapek's *R.U.R.* might result, not from overwork, but from perceived discrimination. Humans form associations of affinity: the distinction between "them" and "us" has been the basis for human culture from long before civilization was invented. Ironically, the final act by a mostly machine cyborg's human soul might be to ally himself with others like him against fully biological humans.

The Twilight Self

The Brain-Body Duality Blues

How will human society treat advanced cyborgs—as something less than human, as bodies with sophisticated life-support devices appended, or as persons with full rights?

Where does the core of a human being reside? Is there a point at which mechanical substitution will be so extensive that an individual will cease to be considered human? And who will decide when and if a cyborg is denied the rights and privileges of personhood?

Taking a step sideways into a different inquiry may yield us some of these answers.

Questions of Life and Death

Medical ethicists have already begun to grapple with such issues, as they consider the related question of when a person can be considered to be dead, in the light of new medical technologies that mimic and support various levels of body and brain functions.

Until advanced medical care became available, the person and the body died at the same time. In old movies, the doctor held a mirror under the nose of the patient—if he fogged it with his breath, he was alive. If not, the patient was buried. Or the doctor took the patient's pulse. No pulse, no life. But we now know that pulseless, nonbreathing, seemingly dead people can be brought back to life by laypeople trained in CPR.

Modern medical equipment can do even more. Bodies with no higher brain activity, in permanent vegetative state (PVS), can be kept alive indefinitely by ventilators and good nursing care. This forces us to reconsider what we mean by "death," and whether there might be more than one death—of the body and of the brain. We are understanding that death is a series of steps, not a single line that, once crossed, is final.

A body in PVS has a dead mind but a live body. Most people consider the person to be dead, although sometimes not the families, who have a harder time letting go of the familiar physical reality of the patient.

For our purposes, the separation of body from brain is significant, as we look toward a future with more sophisticated medical life support systems and implants.

The federal government, and most states, have adopted what might be called the "whole brain" standard of death, after a study in 1981 by a special Presidential Commission. The standard requires that there be no brain waves before the person is considered to be dead. Such is the state of a body in PVS.

Some forward-looking medical ethicists, however, favor a different definition of life, wherein an individual is alive only as long as they can meaningfully interact with others. This could be called a social definition of life, or personhood. This means that once the higher centers of the brain are nonfunctional, the person is gone, although the body may still live without mechanical assistance (of course it cannot feed itself and would soon die). This is known as the neocortical definition of life.

It is named for the neocortex—the "new" cortex, also known as the cerebral cortex. In the folded gray matter of the human brain lie the higher functions like consciousness and thought. Absent or rudimentary in lower vertebrates, the neocortex is maximally developed in humans and in anatomical terms is the most visible difference between humans and other animals.

The neocortex is an association center that receives sensory inputs from the body via the brain stem, and sends instructions back to muscles to act via motor neurons. Distinct from purely instinctual actions (like pulling your hand back from a hot flame), there is time for thought and decision-making in any nerve pathway that is routed through the neocortex. In business terms, it's like asking for instructions from the higher-ups before deciding on a course of action.

From this it is tempting to conclude that the personality resides in the cerebral hemispheres. However, brain physiologists and other researchers would argue that there is no one brain center that contains the personality, or the consciousness. Researchers are finding that even individual memories are not stored in one place, but are distributed among several places in the cerebral cortex.

People who favor the idea of "uploading" consciousness into some other medium, like a future supercomputer, like to describe life in computer terms. The brain is the "hardware," and the consciousness, memories, and the like are "software." In this view, if we could decipher the pattern of electrical activity and read brain waves, we could remove the contents from the brain.

But there is no evidence so far that consciousness can be separated from the structure of the brain. It seems to arise as a consequence of brain organization, as an emergent property: the "wetware" of the brain may be both hardware and software, or the analogy may simply be wrong.

New advances in measuring brain activity are beginning to open up the field of neurophysiology, or brain functioning, and may answer some of these questions.

To look at the brain as a whole, researchers are increasingly using brain scans, magnetic resonance imaging (MRI), as a research tool. Developed as a method of diagnosing problems in patients with head wounds or strokes, the scans are produced by sophisticated electronics that record snapshots of blood flow in the brain. The amount of blood in a particular region of the cortex can reveal where brain activity is focused when we are involved in various tasks. But researchers aren't sure what the images reveal, and in light of the memory research, whether they should expect a tight mapping between a task and a particular area of brain activity. The technology is less than a decade old, and much more research needs to be done to answer these questions.

Neural Prostheses and the Definition of Life

The same characteristics that determine if a person is alive or dead will be used to decide if a cyborg is human or machine.

We currently don't consider someone with multiple replacements (hips, knees, a heart pacemaker, metal pieces here and there substituting for bone, false teeth, plastic eye lenses, hearing aid) to be less than human, but merely unfortunate. None of those parts defines what it means to be human, and the overall structure of the body is still maintained by biological processes.

In the Middle Ages, the heart, not the brain, was considered to be the seat of the soul. In such a worldview, would a person with an artificial heart be human, or a soulless being, a golem?

And what about patients who receive neural implants?

Anticipating more sophisticated medical technology, and the difficult decisions they will usher in, some ethicists favor moving away altogether from biological definitions of life and death.

James J. Hughes, a medical ethicist formerly associated with the University of Chicago, argues that as we look to the future, we must increasingly make a distinction between "biological" and "social" definitions of life and death. Many agree with him. "With the remediation of the brain stem and other body regulating structures, we will be forced to acknowledge that the destruction of the 'integrative' [vegetative] functions of the body is an inadequate definition of death, since the social person will remain intact."

By this definition, if the social person can be kept alive artificially, as a live brain in a nonfunctioning body, he is still a person as long as he is conscious and able to interact with others.

Today, external mechanical devices can keep a body breathing, circulate its blood, and feed it. Future advances in miniaturization might enable these currently bulky machines to be implanted in an accident victim, in effect creating an artificial brain stem.

The brain stem is the lowest part of the brain, located where the spinal cord enters the back of the skull. It controls automatic and "vegetative" body functions such as breathing, heart rate, blood circulation, and blood pressure.

Artificial support to keep the body going is in line with current medical practices. Society has embraced the idea of the heart pacemaker and other medical devices, without which tens of thousands of humans currently functioning as full persons would be dead. After all, people interact with each other's social personas, not with their diseased livers.

Ethical issues will assume more and more importance as the number and sophistication of brain-support and neural implant devices increases. In the future, doctors will be faced with the decision of continuing or withholding treatment depending on whether their patient is a person or not in the eyes of society.

Keeping Connected: Brain-Computer Interface

He was the nearest thing to a dead man on earth.

He was a dead man with a mind that could still think. He knew all the answers that the dead knew and couldn't think about. He could speak for the dead because he was one of them. He was the first of all the soldiers who had died since the beginning of time who still had a brain left to think with. Nobody could dispute with him. Nobody could prove him wrong. Because nobody knew but he.

In the film (and novel) *Johnny Got His Gun,* Dalton Trumbo's main character is a multiple amputee, kept alive by doctors as an experiment because they believe him to be a brain-dead vegetable. As a result of injuries sustained in a World War I battle, the protagonist is limbless, blind, deaf, and unable to speak. He drifts in and out of consciousness, gradually realizing his condition.

With his remaining senses, feeling and vibration sensitivity, he explores his world.

Something very important was happening. He had a new day nurse.

He could tell it the minute the door opened and she began to walk across the room. Her footsteps were light where those of his regular day nurse his old efficient fast-working day nurse were heavy. It took five steps to bring this new one to his bedside. That meant she was shorter than the regular nurse and probably younger too because the very vibration of her footsteps seemed gay and buoyant. It was the first time within his memory that the regular day nurse had not appeared to take care of him.

He lay very still very tense. This was like learning a new secret like opening a new world. Without a moment's hesitation the new nurse threw back his covers. And then like all of the others before her she stood quietly for a moment beside his bed. He knew she was staring down at him. . . .

She put her hand to his forehead and he felt that her hand was young and small and moist. She put her hand to his forehead and he tried to ripple his skin to show her how much he appreciated the way she had done it.

During the course of the novel, he slowly realizes his condition, and struggles to communicate. He begins signaling in Morse code by tapping his head against the pillow.

While he tapped he was praying in his heart. He had never paid much attention to praying before but now he was doing it saying oh please god make her understand what I'm trying to tell her. I've been

alone so long god I've been here for years and years suffocating
smothering dead while alive like a man who has been buried in a cas-
ket deep in the ground and awakens and screams I'm alive I'm alive
I'm alive let me out open the lid dig away the dirt please merciful
christ help me only there's no one to hear him and he's dead.

He is ultimately successful, but in the novel, society in the 1920s
couldn't admit that he could still have a life. In the end, he is drugged
and locked away in a secure wing of an army hospital, condemned to
dream away his existence.

The film and novel are fiction, but many people—victims of
spinal cord accidents (ten thousand per year in the United States),
brain stem strokes or diseases such as Lou Gehrig's disease (the popular name for amyotropic lateral sclerosis)—are trapped within partly
or wholly unresponsive bodies. High-profile examples of active minds
within damaged bodies, like physicist Stephen Hawking and the de-
ceased actor Christopher Reeve, have changed profoundly how we
look at these people, and expanded their prospects for a fulfilling life.

Medical researchers teamed with electronics wizards are reaching out to help these people. Experimental implanted devices allow
paralyzed patients to communicate by moving a cursor on a computer screen—with only their brain waves. These are known as
"brain computer interface" devices, or BCI.

This technique is being used to help quadriplegics: people
with injuries high up in the spinal cord, or higher up, in the brain
stem.

Dr. Philip Kennedy, of Emory University School of Medicine in
Atlanta, likes to call this technology "a mental mouse." Teamed with
neurosurgeon Dr. Roy Bakay, Kennedy inserts a tiny glass cone containing two tiny gold electrodes and an electronic chip into the motor cortex of the brain. This is the neurotrophic (nerve-feeding)
electrode device. From the motor cortex arise the signals for the
nerves of the muscles we consciously control. Among them are arm
and leg muscles.

Enriched with chemicals that entice nerve cells to grow toward
them, the electrodes are soon in contact with living brain cells that

have invaded the glass cone. The paralyzed patient thinks about moving the cursor on the screen, and brain cells in the motor cortex are activated. When they fire, their electrical signals are picked up by the electrodes and transmitted by the chip to a receiver and amplifier. The system is powered by an induction coil placed over the scalp.

The brain waves are transmitted by a computer to power a cursor the way a computer mouse does.

"The trick is teaching the patient to control the strength and pattern of the electrical impulses being produced in the brain," according to Dr. Bakay. "After some training, they are able to 'will' a cursor to move and then stop on a specific point on the computer screen. If you can move the cursor, you can stop on certain icons, send e-mail, turn on or off a light."

After learning to control the movement of the cursor by thinking about it, the patient can move it to icons that represent phrases, or letters of the alphabet. In this way a completely paralyzed patient, who cannot otherwise communicate, can again interact with the environment and with people.

This last is crucial: patients who can communicate through implants or other techniques live longer than those who cannot. Humans are not loners, and the need to communicate is paramount.

Although still experimental, the BCI technology is successful enough that research continues on improvements. The combination of BCI technology with advanced life support dramatically improves the quality of life for individuals with upper spinal cord and brain stem damage.

Several other groups of researchers are working on the problem of translating brain cell signals into movement, using computer models to help them. Recently, a group at Duke University reported success using owl monkey brain signals to move an artificial arm.

Their first hurdle was to find the area in the brain that controls the monkey's flesh-and-blood arm. When the monkey moved its arm, signals emitted by the brain cells in that area were recorded and analyzed. The computer then sent the chosen signals to the artificial arm. If it mimicked the monkey's arm movement, they were successful.

The difficulties should not be underestimated. Most things about the operation of the brain are still not understood. In the owl monkey (and presumably the human) cerebral cortex, the Duke team had to choose in which of several possible areas to implant the electrodes. Dr. Johan Wessberg's team tried five of them before finding one that worked, and it was not the one they had expected.

Then, electrical signals from the brain cells had to be deciphered by computer models to isolate just those needed to move the artificial arm. Eavesdropping on groups of brain cells has only recently been attempted, and scientists are in the early stages of understanding what the signals mean. Do not expect to see this technology made useful for people very soon.

Horizons

Our engineering of the human body extends down to the cellular level. Scientists at the University of Kentucky's Center for Sensor Technology are working on tiny implantable sensors, like microelectrodes, to monitor the chemical communication between individual brain cells. Sensors detecting these tiny, ultra-rapid chemical changes relay the information to external computers, allowing the scientists to monitor brain function without disturbing it.

The microelectrode research is conducted on animal models like rats and lobsters, but the ultimate goal is to understand the signaling between nerve cells in the human brain. These interactions break down in degenerative diseases like Parkinson's, Alzheimer's, and Huntington's diseases, causing a loss of brain function and control. Chips aren't the only brain implants being used. Other devices control epileptic seizures, chronic back pain, certain stomach problems, and tremors of Parkinson's disease with electric pulses delivered directly to the brain or nervous system. Cardiac pacemakers, successfully implanted since 1958 in hundreds of thousands of individuals, have inspired a range of similar devices for different conditions.

Parkinson's disease is an extremely common neurological disorder affecting over one million people in the United States. The Activa Tremor Control System (Medtronic, Inc.) is an implanted medical device, similar to a cardiac pacemaker, that uses mild electrical stimulation to block the brain signals that cause persistent tremor (severe shaking) in the hands, legs, or head. An electrode with four leads implanted near targeted cells in the thalamus (the brain's message relay center) jams the neural network to interrupt the tremor. The electrodes are connected to a neurostimulator implanted near the collarbone. The stimulator is turned on by waving a magnetic wand over it.

Adapted cardiac pacemaker technology is also used to stimulate the spinal cord in tens of thousands of people with otherwise untreatable chronic pain in their legs and back. Vagus Nerve Stimulation (Cyberonics, Inc.), or VNS, uses similar technology to control otherwise intractable epileptic seizures. To date, over eight thousand patients of all ages with a variety of seizure types have been treated by VNS in the United States and Europe.

This technique applies electrical signals to the vagus nerve via an electrode implanted in the neck, and transmitted upward to the brain. The low-level electrical impulses somehow interfere with the electrical storm in the brain that results in an epileptic seizure.

The great success of cardiac pacemaker technology was just the beginning of what may be a huge industry in brain-implant devices.

Personhood Beyond the Body

In time, we will be able to remediate closer and closer to the cortex, and even into the cortex itself. James J. Hughes asks, "How much of one's motor skills, memory and cognition may one lose to be treated as dead, 'socially dead' or 'sick enough to not require further medical treatment or feeding,' *if those abilities can eventually be restored?*"

These may eventually be applied to cerebral tissue. Ultimately, argues ethicist John Hughes, a true definition of personhood should transcend the purely physical. The "social personhood" concept asserts that citizenship, rights, and value adhere not to bodies, but to "subjective persons."

> Once we begin to remediate cerebral cortex injuries I believe we will
> be forced beyond a neo-cortical definition of death to one focused on
> the continuity of subjective self-awareness. Those who have a contin-
> uous sense of self-awareness, *in whatever media,* will be considered
> social persons, with attendant rights and obligations.

In this definition of life lie the seeds of the coming issues about advanced cyborgs. Suppose society moves away from the body-as-person concept. If we accept social identity alone, that could lead to granting personhood status to uploaded human consciousnesses, to brains maintained outside of bodies (or transplanted into synthetic bodies), and then to artificial persons entirely.

That large-scale upheaval in social values is unlikely soon, but the entry of ethicists into the debate shows that it is no longer simply a story idea for fiction writers.

Is a brain in a metal body still human? Science fiction authors who have treated the issue opt for characters with deep psychological problems, but where's the story about boringly normal ones (that is, success in ordinary terms)?

As such, it was the subject of stories, novels (William Gibson's *Neuromancer*), and not a few B movies (like *Donovan's Brain*). Damon Knight's 1968 short story, "Masks," discussed earlier, explores what he terms total prosthesis, the transplant of a brain into a wholly artificial body—the extreme cyborg. Certainly this phenomenon would bring unforeseen personality changes, because no psychologist doubts the necessity of bodies to our mental stability.

One conclusion is clear: Take away completely our connections to the external world, and we go crazy. People in immersion tanks—where they float weightless in warm saline solutions, insulated from light and sound—soon start imagining stimuli. With no signals coming in, their minds turn up the gain, trying to pull some signal out of the blankness. Soon enough, they hallucinate, seeing and feeling inputs that do not exist. Something like this happens to those living in deserts, where the lack of discernible features calls the mind to summon up its inner resources. It is not an accident that religious mystics, like John the Baptist, go alone into the desert's featureless retreats for enlightenment. A cyborg whose body is distant and divorced from much of the customary human constellation of sensations and emotions, as in "Masks," will be hard to identify as human: "No more adrenal glands to pump adrenaline into his blood, so he could not feel fright or rage. They had released him from all that—love, hate, the whole sloppy mess—"

So transplanted people do not just lose some feelings, but rather, as discussed earlier, they experience new ones. This sensitivity seems probable for all stages of the cyborg, as new states of the human condition.

Technodreams

The ultimate promise of the cyborg is better people. Yet that means redefining personhood in ways that will seem radical to many. Not radical now, at the early stages—but eventually, perhaps in half a century, cyborgs may be a significant social issue.

Public controversies frequently pit people talking statistics against those talking myth. As futurist Walter Truett Anderson puts it, "the rationalists with their hard disks full of economic or scientific information bump against invocations of Frankenstein and Gaia."

The old treatment modes—preventive, palliative, and curative—shall soon give way to a powerful fourth: substitutive. People find unremarkable an aging athlete with an artificial left shoulder, wired back together after a softball accident. Soon he or she may need a pacemaker, or even some of the odder additions people accept: artificial sphincters, prostheses, cochlear implants to restore hearing. Mechanical, they seem as natural to us now as eyeglasses and tooth fillings.

Anderson predicts that the next major augmentation will probably be organ transplants with artificial assists, both through drugs and in-body cyborg devices. This will bring, he says, a wholly "new chapter in the history of animal husbandry—and indeed in the history of life on Earth—because there has never been an animal able to exchange entire organs with those of other species."

Human–human transplants are commonplace, with new anti-rejection drugs and better surgery spurring their survival. In the last five years costs have been cut nearly in half, so that a kidney transplant now costs fifty thousand dollars, and a liver two hundred thousand dollars.

But with transplant numbers rising at 50 percent every six years, donors are scarce. China routinely executes thousands of criminals a year on the operating table, so their organs can be "harvested" optimally. At the top of the list for recipients are those with political or economic clout. The Chinese run a thriving business selling organ transplants to rich foreigners, too—a situation predicted in 1968 by Larry Niven's "The Jigsaw Man."

Even in China, the organ supply problem is overwhelming. Many scientists and doctors now pursue replacements from animals. Pigs have organs the right size for humans, and such "transgenic" animals will be used instead, possibly with artificial devices to help. Genetically engineered with human proteins to cloak offending pig molecules, pig organs will fend off our defenses, reducing the rejection problems. The key development is information at the molecular level.

Where should we let it drive us? No moral anchors here seem trusty. Invoking Nature with its implied supremacy ignores that many cultures have fundamentally differing ideas of even what Nature is, much less how it should work.

Other cultural guidelines—religious doctrine, scientific objectivity, fashion—are similarly mutable and local, necessary perhaps but not sufficient as guides. Anderson's "blessing and scourge of our time" is the dizzying multitude of our options. Cultures must clash when the questions are greater than regional.

Who will win, in this future? Our world society has many different cultures, so some may ban technologies while others spur them on, if only to market them for profit. Here lie the seeds of international strife, all centered by ideas about the limits on defining human.

The nature of technodreams is to force us to ask ever more fundamental questions.

i i

Robots Plus

Anthrobots—Machines R Us

In L. Frank Baum's classic, *The Wonderful Wizard of Oz*, the Tin Man became what he was by being careless. He was a woodsman with a sharp ax, and apparently overeager. First he accidentally chopped off a hand, then an arm, a leg, and on to other parts. Each time he got a tin replacement, but his clumsiness apparently did not discourage him from continuing to chop wood. Eventually he was wholly tin, and when he enters the story he needs a heart.

So he is a robot built on a human mold, and like so many fictional 'bots, longs to be at least partially human. This last assumption was old long before *The Wonderful Wizard of Oz*, as if anything that resembled us would *of course* want to be like us.

We all know this common image of robots as pseudo-humans. Hollywood's many images are mired in the science fiction of more than half a century ago. Spielberg and Kubrick's 2001 *A.I.* emerged from a 1969 Brian Aldiss short story. Isaac Asimov's 1940s stories came to the screen in *Bicentennial Man* (1999) and *I, Robot* (2004). In 2005's *Stealth* the AI is basically the HAL of the 1968 film, *2001*, with a curt personality and lots of real cool weaponry. The movies alerted us to the prospect that if a future robot sounds like HAL in *2001* it's either low on batteries or planning to attack us.

The old idea of robot cars is working out, but slowly. The DARPA (Defense Advanced Research Projects Agency) Grand Challenge holds a robot vehicle race every year. The 2004 event echoed earlier years—the entries made it a few miles across desert, then stalled out, crashed, or self-destructed. Though little progress came from years in this spirited competition, despite considerable funding, the 2005 race suddenly produced four finishers, and Stanford University won the $2 million prize. The rough terrain took much innovation to cross. In the 1990s, cars from Carnegie Mellon drove themselves to Washington, D.C., and back, but highways are far simpler to navigate than open deserts. It may take another decade before we first see cars passing us with no drivers. Indeed, they will probably have darkened windows, not to astonish nearby drivers.

Films especially inflate our expectations. It's not hard to have a human in a metal suit do fantastic things, so we expect that in life. These hopes led in the mid-1980s through the 1990s to the "AI winter" when funding for the field dried up. It has recovered in the last decade, driven both by defense applications and by burgeoning commercial products.

How far could such development go? Can a robot have even a spiritual function? If they could be made to be qualitatively different from ordinary humans, say with some appetites and motivations altered or erased, why not?

In Anthony Boucher's 1950s short story, "The Quest for Saint Aquin," the saint turns out to be a robot. Intelligence in silicon has finally advanced to a perception of the holy, and the humble robot is the Word Made Steel. This hope that by making a thing of metal we can make a humanoid being that rises above the desires and constraints of the body comes from a long tradition in religions of seeking to abandon the human body. Rather than becoming a spirit, perhaps it could be better to remain made of matter, yet improved?

• • •

Such ideas are threatening to many people. Robots are in their infancy, but already (especially in Western cultures) call forth images of humanity's eventual doom. Even science fiction that embraces technology seems uneasy about robotics. The cyberpunk subgenre, now largely mined out, rejected the body in a way that reminds many of Puritan instincts. We Westerners carry cultural baggage that may well inhibit our development of such technologies, compared with, especially, the Japanese.

Part II explores some real robots and ideas about them. It is far more probable that the next decade or two will give us "niche intelligences" that filter spam, carry items in hospitals, or clean our houses, rather than affordable, humanoid robot slaves that jump to do our bidding. Some science fiction writers like Michael Flynn and Robert Forward have treated this humorously, with "artificial stupids" that bumble through their jobs, but are still cheaper than humans. After all, the hardest skills for robots to develop are those we learn as children—throwing balls and catching them; balancing on narrow planks. We do these on playgrounds for fun, while MIT spends millions to teach machines how to do them badly.

Even these task-specific devices will seem remarkable, even bizarre—but we will get used to them. Hiroaki Kitano, a prominent Japanese robot engineer, predicts that half a century from now, humans and robots will play soccer on even terms in the World Robocup. He believes that robots in competition will be unremarkable by then, but even he does not suggest that the teams may be mixed human and robot. That may say something about our times and their anxieties.

Quite probably, the hospital 'bots just arriving may make us think of them more benignly. These are specialized therapy 'bots, built by a team at MIT to make up a "gym" where stroke victims visit a series of stations to get smart feedback as they relearn how to use their arms and legs. Patient, with perfect memory of prior patient performance, they can guide a stroke victim's leg through up to 1,500 repetitions in an hour, without getting bored or impatient. Many of the exercises are games, like the wrist trainer that wraps an injured hand, letting the patient try to get a ball to roll into a hole in the middle of a swaying table. Interactive Motion Technologies, Inc., a small company designing these, has had good clinical success.

Patients like the steady attention and focus of the machines, which look a lot like ordinary exercise apparatus. Such acceptance may lead us to accept robots in our bodily world, as helpers, not opponents.

CHAPTER 5

Robots to Order

Here comes the Robo-[fill in your need here].

Robby, the robot of Fred M. Wilcox's film *Forbidden Planet*, synthesized food and clothing, drove a car, and ran a complex household. Marvin, the paranoid android of Douglas Adams's *The Hitchhiker's Guide to the Galaxy*, complained about having to do menial chores with "a brain the size of a planet." Jenkins, the eternal robot of Clifford Simak's *City*, gently guided the demise of human and rise of canine civilizations in the solar system. R. (for "robot") Daneel Olivaw ran an entire galaxy in Isaac Asimov's Foundation series.

Are these the creatures we imagine will want to take care of our domestic chores? Why spend the money on machines that could do just as well with an IQ of 60? Engineers have long known this truth—don't overdesign. Something a bit simpler seems appropriate, and is now just arriving in the marketplace.

Robot Menials in the Home

Industrial robots had their start in 1961. The first, called Unimate by its inventor, George Devol, and Devol's partner, Joseph Engelberger, was installed in an automotive plant owned by General Motors. Although they have since become a staple of the automobile assembly line, industrial robots are not a huge business. The market is worth $4 billion plus worldwide, compared with the market for teddy bears at $1 billion annually, and General Electric Company, which made $117 billion in 1999.

Although dreamed of by the general public since the 1939 World's Fair, robots for home use are still not generally available. In a Ray Bradbury 1940s short story, the house of the future is fully interactive. Lights turn on and off on schedule, the kitchen prepares meals and announces when the food is ready, a robo-vac detects soil and cleans it up, and a robo-butler serves drinks, sets a fire in the fireplace, and lights a pipe for the smoker.

In Bradbury's story, the "friendly" house technology is used as counterpoint to the neutron bomb that kills the people and their pets, leaving the house with no one to care for. Bill Gates has a large, heavily computerized home that sports embedded intelligences. Rooms respond to a person's desired lighting level, air scent, temperature, and even the "views" on flat wall screens, when they sense the person entering the room. (Gatherings of several people demand some compromise, but allegedly Gates's preferences rule when he or his wife are present.)

But a house filled with humanoid robo-butlers and maids is still far in the future (and, one hopes, also without the bomb blast). If it's ever to happen, chances are much of the innovation will come from MIT's Media Lab, breeding ground for new ways to design and use technology. There a Counter Intelligence project—kitchen counter, that is—was set up in 1998 to bring "smart" technology into the kitchen. They're exploring devices like a microwave with an electronic reader to automatically recognize what food is put into it and how long it needs to be cooked. That's actually not much of a step forward from the 1980s heat probes in conventional ovens and even some microwaves. Designed to be inserted into a piece of meat or a casserole, they turn off the oven at a predetermined temperature.

Another device being explored is an interactive recipe countertop, with audio and video to walk the chef through the instructions.

Then there's the automatic spatula with different adjustments for different tasks, a temperature-sensitive oven mitt that talks, and a knife that detects bacteria.

There may be enough demand to make these commercial. Still, what kind of help do people really want in their kitchens? We already have dishwashers with variable washing times depending on the amount of soil in the rinse water, and clothes dryers that sense the dampness of the clothes tumbling within them. Coffeemakers, dishwashers, and ovens can be programmed to go on and off at preset times. All of those can be considered to be smart to a degree, but they will not be at all versatile, so probably they properly will be seen as mere tools.

Managers are now pondering what clever appliances will have a market. Here is where intuition comes to bear. What about a refrigerator that tracks the age of mysterious frozen packages and half-forgotten leftovers? Odor sensors in the lids of dishes could warn of bacterial growth inside. Sensors in the fridge would detect the warnings and post them on a door display, or simply tell the next person who opens the door looking for a snack. Fridges with built-in bar-code scanners could easily track inventory in the home as well as in the grocery store. Link the fridge to a personal computer with Internet access, and it could dial up Webvan.com and place your order. Alas, Webvan.com is now defunct, but may be replaced.

Going to be out of town for a few days? Forget about preprogramming your VCR to record favorite shows. ReplayTV, a smart device already on the market, searches the cable for programs you like and records them on a hard drive. This is the beginning of integrating the entire house to run itself without our constant instructions.

In future, how will we control all this? The next big advances will have to be about coordinating and networking within the house. Nicholas Negroponte, past director of the MIT Media Lab, has a vision of all these smart devices working together to make life easier. One small example: between the alarm clock and the coffeemaker, you awaken to the smell of fresh-brewed coffee each morning, even if it's not at the same time every day. If the house is Internet-connected, traffic reports could be scanned, and the wake-up time adjusted accordingly. These services, performed by servants for the ultra-wealthy of today, might fairly soon be within reach of a much wider slice of the population.

Slowly, we see the true house 'bot emerging. In 2005, the prospects seemed good.

Where Are You, Wakamaru?

A striking contrast emerges between the two leading cultures producing robots: the Japanese and the West (mostly the USA). Japan has a strong xenophobic strain, hostile to immigrants; the United States does not. The Japanese welcome humanoid robots, whereas in the United States they are the classic threatening science fiction figures—though R2-D2 is a counterexample. This has led to the development of many humanoid robots in Japan, yet few in the States.

With its humanoid Wakamaru robot, Mitsubishi aims to create a robot that can sustain meaningful relationships with human beings, initiating conversations with family members and offering services such as alarm, news, weather, and e-mail dictation. Wakamaru can look after the house, provide video streams over cellular networks, and cull useful information over the Internet, while maintaining its own autonomous "rhythm of life," according to Mitsubishi. Wakamaru was designed by Toshiyuki Kita, who patterned the robot after a growing child. "Wakamaru" comes from the childhood nickname of Minamotono Yoshitsune, a twelfth-century Japanese samurai who engineered military victories that enabled his brother Yoritomo to gain control of Japan. The name is associated with growth and development.

Wakamaru can identify up to ten people by their faces, including two it considers "owners." Using speech recognition technology to identify ten thousand Japanese words, Wakamaru's speech synthesis capabilities include voice modulation and using gestures when speaking. It even recognizes nicknames given to it by users, Mitsubishi says.

A panoramic top-of-head camera enables Wakamaru to find its position in the house, according to the ceiling. This camera also allows the robot to face others when speaking to them or being spoken to. Unthreatening Wakamaru stands just shy of four feet tall (100 cm), and weighs sixty-six pounds (30 kg). It can travel at one kilometer per hour, avoiding objects and identifying moving people. Its claimed battery life is two hours, after which the robot returns to its charging station before power fails completely. It maintains Internet access and communications capabilities while charging, too.

But while it can go up to one centimeter high from the floor, getting over minor obstacles, it can't climb the stairs yet (as of 2005). It responds to voice, and it's not designed to be used for nursing care, because it lacks muscular and grip strength. Therefore, it's not equipped for lifting and nursing assistance. It's a faithful, polite butler, not a nurse.

Another approach is Nuvo, a general-use robot made in Japan, sold for about six thousand dollars. As Japan ages, its elderly may need simple companionship, and Nuvo can provide empathy. Only fifteen inches high but sporting fifteen motors, it can be a baby watcher, security patrol, and general helpmate, able to use the Internet and report back on one's apartment. If it falls, it gets right back up. It responds to orders given in a conversational tone, and can casually report on news and e-mails it gets from the Internet. As a helper it uses gestures, handing over a tool when asked.

One who kept company with Nuvo for four days said, "Much of the time it felt like having a dog around, without my having to feed it." Perhaps Nuvo is the first of the true service 'bots. It calls forth easy roles—intermediate between pets and real humans, and so easy to anthropomorphize—with its limited menus of response and nuance, both verbal and physical.

But suppose Wakamaru or Nuvo get too smart, too savvy?

Privacy and the Servant Problem

The downside is obvious. A linked world is also a nosy one; we can expect robots to be no different. Indeed, tiny 'bots slipping into a room to listen and see unnoticed will be a common method used by private detectives, commercial espionagers, and nations. A personal butler can be a good thing, but has to know a lot about you to be effective. The same goes for robo-servants and smart, networked appliances. We will in the future pay to keep our privacy, not to let our clever surroundings obtain too much information about us. Your TV or home climate system does not have to know who all your guests are, just a few details about their preferences. Anything more, and they could be hijacked by paparazzi, nosy neighbors, or worse.

Although talking devices are a cliché of stories and movies set in the future, many people don't want their machines to talk. Elevators that announce upcoming floors are useful, but do you really want the car to nag you about putting on that seat belt? Most people did not, so that feature disappeared. Also gone is the alarm system that warned innocent passersby when they were "too close" to the outside of the car. That presumptuous message probably accounted for lots of dings by enraged pedestrians. Gone is the talking camera that told the user if the shutter speed was wrong ("too dark") or the focus was incorrect ("too close") in tinny, Japanese-accented English. Visitors to the Media Lab similarly nixed the microwave oven that announced when the food was ready. A quiet beep was quite enough, thanks.

Recall that *Star Trek* movie when Scotty addressed a computer in the 1980s with instructions, and was surprised that it did not respond? Voice activation of computers is now being introduced again for the PC market with great fanfare—but actually, Apple tried and then discontinued the feature in the 1980s. After a short training period to allow the computer to recognize the user's voice patterns, the computer would execute any commands in the pull-down menus on demand. All that remains in today's software is a system of spoken "alerts" that trigger if the user doesn't respond quickly enough to on-screen warning messages. Cell phones now dial people in your address book by voice command, but systems that can give layered information are just on the horizon. Many telephone interactions, such as airline reservations, are computer-handled with voice recognition that simulates an encounter with a real person fairly well.

The next decade will see an emerging class of small, distinctly nonhumanoid robots. Robotic pool-cleaners were the first of the breed, roving underwater vacuums resembling turtles that slowly creep around backyard pools, sucking leaves and dirt from the concrete pool bottoms. Behind them trails a long hose that delivers the dirt to the pool filter. They are fairly common in Europe as well as the United States, but interestingly are marketed as being automatic, not robotic.

So were the next ones to be introduced as the ultimate in luxury for suburban homeowners—robotic lawn mowers. Although inventors have experimented with automatic lawn mowers for fifty years, in 2000 two manufacturers released "mow-bots" in the United States that had been available for a couple of years in Sweden. At least one other is scheduled to make an appearance shortly.

Early models were too dangerous, too costly, or never quite made the grade in lawn grooming. Due to considerations of safety and control, existing models can be operated on a lawn only within a wired perimeter. They also have lift and tilt sensors that stop the cutting blades if the mower is disturbed. As a further safety precaution, the blades rotate only when the mower is moving forward. Some are solar-powered, while others use rechargeable batteries. They are quiet and slow, subtle additions to the landscape.

Like the pool cleaners, mow-bots are considered automatic, not robotic. Fair enough, for now—but they are being upgraded as the manufacturers learn from field experience, and computing gets steadily cheaper. Soon enough they will respond to changing conditions, like stopping work if it rains and the grass gets too hard to cut. They will avoid pets and humans, not chop into the sprinkler system, and other things a handyman would do.

For the first time, household devices are being deliberately marketed as robots.

According to one manufacturer, Friendly Robotics, Inc., its mower's wire sensors recognize an "active" perimeter wire (it works like a low-voltage electrical fence) to prevent it from moving outside the desired area. If it is placed outside the perimeter, it stops, eliminating opportunities for robo-mower liberation. The idea of a rogue lawn mower marauding through the neighborhood, mowing down flower beds and menacing pet cats, was probably too much for the legal department of the manufacturer.

We've had toys that dealt with these problems since the 1970s. Touch-sensitive bumpers recognize solid objects like rocks and tree trunks, causing the low, roughly turtle-shaped mower to change directions and move away. Furthermore, running on rechargeable batteries, the mower is soundless and emission-free. At least one of the models is programmed to return to its recharger when the batteries run down.

Robotic household vacuum cleaners came next, best exemplified by the iRobot Roomba, a low-slung device introduced in 2001. These small, battery-run, flat robotic units roam through the house, assessing the level of dirt in their incoming air stream with optical sensors (like some dishwashers do with rinse water). The robovacuums are programmed to move back and forth over a dirty spot until the incoming air is clean. Unlike the pool vacs, they keep the dirt in the mobile unit, can be kept out of areas by a microwave beam "fence," and their search patterns get smarter with each generation. Leave them to roam and leave for work. Roombas hum around, then scoot back into their recharging stations when they need a break. They seem as smart as a cat, but not as lazy. Their iRobot cousins went further—sturdy crawler 'bots crawled into the World Trade Center, searching for survivors. A visit to iRobot's Web site shows specialist 'bots suited for a wary world that wants expendable machines for exploring sewers and sniffing out anthrax.

Manufacturers expect that in a few years a typical North American home could own several robotic devices. An excerpt from the 2001 Friendlyrobotics.com Web site sounds like a perky parody of the 1950s exuberance for household appliances:

Friendly Robotics aims to become a global leader in the emerging sector of Consumer and Home Robotics. Consumer and Home Robotics is the killer app of the Intelligent Appliances market and Friendly Robotics has strategically positioned itself as the leading player in the field. The Company's state of the art and highly effective robotic devices perform mundane, repetitive and time-consuming tasks, thereby liberating people from routine chores and improving the quality of life for users. Friendly Robotics is the first, and currently the only company in the world to focus on Home and Consumer Robotics.

Marketing departments are never ironic, except unintentionally. Perhaps they never heard of the Hitchhiker's Guide Sirius Cybernetics Corporation, which makes "plastic pals that are fun to play with" and endows elevators with "genuine people personalities." The Web site says:

> The RL1000 Robomow will automatically depart at the times you have pre-set, cut your lawn and return to the Charging Station at the end of each operation. It will then recharge and wait for the next scheduled operation. Simply schedule a weekly program and forget about mowing for the entire season!

In the near future, then, humans will continue to cook and scrub for themselves or hire humans to do it for them, but robotic assistance for other chores like pool cleaning, lawn mowing, and floor dusting will be available, for a price. Robo–snow blowers to clear the drive are hard on the heels of the robo-vacuum. Indeed, there may be one or more robots doing domestic drudge work for each of us within a few years.

Robotic golf carts that follow golfers around the course are on the horizon, but don't expect one to help you with your game. "Everybody has seen *Star Wars*," says the product's inventor, Ron Davies. "People ask why our robot won't just hand them the five iron. They want to know why it won't talk. . . . But remember when cars would say things like, 'Your door is ajar,' and everybody hated it? I want to make sure that the technologies we use are what people really want."

So the InteleCady from GolfPro International is strictly a golf club hauler, moving around the course by consulting its internal map and global positioning system (GPS), avoiding objects by sonar sensors. It won't run you over either—a bumper automatically shuts off the motor if it hits something. And when you want it, just sum-mon it by radio signal.

The robots just described are real commercial products, on the market today. In this excerpt from *The Coming,* author Joe Haldeman takes them a step further, imagining yet another use for semi-intelligent robots:

> Dan whistled and pointed to the screen. The large camera rolled up to it and seemed to concentrate.
>
> "Daniel," it said in a soft woman's voice, "please come adjust my raster synchronization."
>
> Dan shook his head. "That's automatic in the new models." He got up and peered through the camera and fiddled with a pair of knobs until the picture of the wallscreen settled down.
>
> He returned to his seat and the small camera climbed up onto Bell's desk and stared at her. She looked at it warily. "Am I supposed to talk to it?"

Can we expect political fallout from robo-appliances? Inevitably, people in entry-level jobs will be displaced by the new machines. Pool cleaners, golf caddies, and high schoolers hoping to cut lawns or shovel snow will not be needed; demand for unskilled illegal immigrants will ebb.

No tragedies here, but . . . at every step of the way, the mecha-nization of labor has been seen by some as antihuman. Thirty years ago, garment unions decried automatic shirt-cutting machines. With the advent of computers, other routine jobs have been eliminated or reduced, like bank tellers by ATMs and switchboard operators by those maddening automated systems.

These household robots are yet another step along the path of freeing humans from routine and repetitive jobs. Yet because they are also a small step along the path toward independent act-ing machines, they may attract more than their share of nega-tive attention. Predictably, within a decade or two no one will miss the jobs they took over. The buggy whip manufacturers are long gone.

What about jobs that no person can now do *at all*?

Gecko, the Wall-Climbing Robot

The Defense Advanced Research Projects Agency (DARPA) calls it part of the Controlled Biological Systems Program, and iRobotics Corporation terms its new beast the Component Technologies for Climbing device, but everybody really calls it "Gecko."

It's a small autonomous wall-walker, mimicking biology. Small creatures are often more mobile than their larger counterparts; geckos can crawl up and over almost any obstacle. Traditional robots can't— they overcome bigger obstacles only by using ever-larger machines.

A biologically inspired gecko-robo may be able to traverse any terrain. Undeterred by impediments many times its own size, it could easily conquer obstacles that would paralyze a typical half-track microrover, turning the present two-dimensional world of rovers into a new three-dimensional universe.

Collaborating with researchers at UC Berkeley's Poly-PEDAL Lab, IS Robotics is currently developing the legs for such a robot, the success of which could take us a few steps closer to realizing this dream. To investigate some of the various climbing mechanisms found in nature, the Poly-PEDAL Lab is studying the methods of many climbing insects, lizards, and amphibians, hoping to find the key to robotic climbing.

For these researchers, real geckos are especially interesting. This lizard is able to cling upside down and scale nearly any vertical surface. With a toe spread of more than 180 degrees, claws, and an elaborate foot pad structure covered with microscopic suction cups, the gecko sticks to surfaces through what some believe to be intermolecular attraction.

This attachment method combined with the flexibility of its spine allows it to effortlessly transition from horizontal to vertical surfaces. Spinal flexion, however, is not essential to climbing. Insects such as the common cockroach are able to transition even with a rigid exoskeleton. These creatures employ additional degrees of freedom in the forelimbs and a yaw of the body to accomplish the same task. The task of constructing similar mechanisms in robots drives the Gecko project.

Gecko's final design remains undetermined, but it will have multiple legs, each with some attachment mechanism, plus a separate, bigger device to make sure Gecko doesn't fall. The legs will each have several degrees of freedom, so it can turn and move between surfaces, still keeping contact with the wall—no matter how sticky or smooth that wall may be. It will copy nature, using claws for softer materials and adhesives for firmer or smoother walls.

The entire robot will be self-contained, carrying onboard power and electronics. It will carry a behavior control macropackage, allowing programmers to easily delegate routine robot behaviors. Sensor technology will allow Gecko to move on its own in an unstructured environment, versatile and situation-smart. Its designers dream of races against real geckos, someday.

For what use? Light and portable, easy to deploy, Gecko could sneak around unnoticed in covert military applications, or generally any place where watching unobserved would be handy. Operating models could manipulate and reposition sensors, carry out reconnaissance and surveillance, infiltrate, and do mine detection.

Gecko robots could venture to handle hazardous materials, inspect dangerous or hard-to-reach places—pipes, tunnels, ventilation shafts, wall spaces—and fetch forth lost items. A future with such robots in it will make people look twice before swatting a "bug."

I See You

The iRobot-LE is not just another Web cam. Already, mass access to immense bandwidth has made it possible to put personal Web video cameras everywhere. The Internet has millions of them. While this proliferation is, in itself, neither good nor bad, it could make privacy a thing of the past. (For implications of this, see David Brin's prophetic *The Transparent Society: Will Technology Force Us to Choose between Privacy and Freedom?* which in 1998 foresaw much of our present and likely future.)

But robotics changes this concern about unwelcome surveillance. What if you own the camera, and it moves and can carry out tasks as well? Suppose you wanted to be able to see what was going on all over your house, especially when you were not there. Presently, you would have to install and wire cameras in every room. That's a lot of cameras, and for a family, it means never knowing if you are being watched or not by . . . who? Perhaps by Mom and Dad.

So use the iRobot-LE, which can be driven, over the Internet, to anywhere in your home. If you want privacy, close your door— that's it. The iRobot-LE ("committed to building robots for the real world") is an example of technology used responsibly to simplify our lives, not make them more stressful.

The iRobot-LE has a head on a long neck, so it can look upward and interact with people in its line of sight. With hundreds of phrases for ritual response and direct audio feed, it functions as an alter-self of the operator. Plainly this is only the beginning of a generation of mobile robots that can serve as stand-ins of varying ability and perception, abilities that will steadily improve as chips get cheaper and servos more supple.

Until now, home robots have been little more than toys, substitute pet dogs and the like. People will cuddle and talk to inanimate objects, as everyone who has a teddy bear knows. The iRobot-LE is different, with sensors that allow it to understand the world around it. It can bring the power of the Internet out of the study and into anywhere in the world where there is an iRobot-LE you are authorized to use. Microwave beams from the computer instruct the mobile home-bot, carrying audio and visual feeds in real time. At a price of about six thousand dollars, it is a high-range appliance, the first of many. Some uses:

Home Sentry: Keep an eye on your home while you're away on business. Outsmart thieves by leaving a set of roving eyes.
Pet Care: The pooch stuck at home? If you're a pet owner, you can empathize with how difficult it is to leave your pet behind. A couple of times per day, wouldn't it be nice to check on how your best friend is doing? If the robot carries a voice emulator, it can talk to your pet in your own recorded voice, soothing it. Eventually, robots may be able to bond with pets and provide companionship with an emotional flavoring for them.

Check on Babysitters: Communicate with your babysitter and see
directly that your children are receiving proper care. And
while you're at it, say "Hi" to them all.

Vacation Home Monitoring: You want to know how your vacation home fared after a big storm ripped through the area. You
want to be sure the water pipe didn't burst during the last
cold snap. You want to verify that there are no signs of break-
in. You can check on all these things with your iRobot-LE,
saving yourself the travel time.

Invite Relatives and Friends In: At family reunions, there are always a few who can't make it. Let your faraway relatives log
in over the Web to participate actively in the festivities.

Monitor Contractors: See how your remodeling is going and ask
questions from a distance.

Elder Care: Visit by proxy with elderly or housebound relatives.
Converse with them and keep them company.

One can readily think of misuses of this technology. If it can be
hacked over the Internet, havoc will follow. But as the iRobot-LE
gets smarter, with better software and faster chips, it will be able to
defend against outsiders. As always, this will be an arms race without end.

Though the iRobot Corporation has an edge with its present
product, there seems to be a wide market for many competitors. For
example, for true technophiles they have marketed the My Real
Baby doll—a child-centered, artificially intelligent, emotionally ex-
pressive, emotionally responsive, robotic baby doll. It (or she)
boasts a web of state-of-the-art sensors and a sophisticated, sensitive software system. She laughs, cries, blinks, breathes, burps, has a
wide range of facial expressions and emotional responses. She responds to how she is played with and how she is held—naturally,
in emotionally appropriate ways, "intelligently—just like a real
baby." At a price of about one hundred dollars, it is an affordable
toy, no longer exotic.

Though designed with children aged three and up in mind,
people of all ages love to play with her. It appears easy for many to
attribute human qualities to robots even though they know the experience is artificial. Probably this will apply to all robots that look

like pets, babies, or other objects of affection. The film *A.I.* brought all these instincts to the fore, with deeper questions of how we should feel about such attachments. In a way this differs little from the teddy bear affections of our childhoods, but future teddies will be better, more responsive, and so harder to let go. There are innumerable stories to come from this, no doubt. People will develop manias for products that have special meaning for them. We may see robot fetishes, as the *A.I.* robotic male prostitute Gigolo Joe (played by Jude Law) made clear. With superior physical attractions and software that made it always say the right, romantic phrases, Joe would be a formidable opponent in the mating game.

A tide of similar products is about to engulf the developed world. They will ape easy human signals, using sight and sound, gesture and nuance to tag themselves as acceptable members of our community—pseudos of our pets, children, neighbors, friends, servants, maybe even lovers. There is ample room here for both farce and tragedy.

Product lines will differ in their sublety, their servitude, their intelligence and nonverbal abilities—all of which will improve with time. As generations of robots come along, they will probably be labeled by the digital indices familiar from software: a RoboButler 1.2 will give way to the snazzy 2.0 and 3.5. Eventually, they will just be called Jeeves, and accepted.

But this engagement with our machines will not have any certain end. Driven by technology, the issues will multiply. We have a great ability to identify with things—some of us name our cars, as if they were pets—and this tendency will become more powerful once things echo our needs.

Eventually, we will see ourselves in symbiosis with machines in a way difficult to glimpse today. But then, this is, in a way, natural. We gained much from being able to domesticate animals, engaging them in genuinely emotional ways, as with our dogs, cats, horses, and the like. The robots will take this much farther.

CHAPTER **6**

Robowar

Most likely, robots will make our battlefields
less bloody . . . for some.

In 1994 Michael Thorpe, a former model maker at George Lucas's Industrial Light & Magic, began public, live robot fights in San Francisco. These Robot Wars began as displays of engineering craft and imagination, allowing the geek community of the area to show off its inventive, destructive talents. Most of the gladiators looked like moving junk piles, springing clever knives, hammers, spikes, electrical arcs, and other instruments of mayhem upon their opponents.

Thorpe quickly drew a large audience. Hundreds of techno-nerds proved quite willing to spend thousands of dollars and hours of labor to make combatants that they hoped might survive for a few minutes in the ring.

A few of the battlers looked benign, but even that was a disguise. A fourteen-year-old girl brought a ladybug-looking robot whose pretty red shell lifted to deploy a hook, which then skewered rivals. Their very names aimed at intimidation—Toecrusher, Mauler, The Hammer, Stiletto.

Their audience grew steadily until a legal dispute closed the games, but not before a promoter saw potential in the spectacle. Similar contests went through a brief pay-per-view series, then ended in 2000 with a slot on TV's Comedy Central. There, battling robots showed their dual nature—focus for malicious mayhem, plus inadvertent comedy. They offer ritual violence directed by their creators using remote control, so they are only the simplest sort of the robot species, incapable of independent thinking and action.

These are techno versions of aggression, a weird blend of the "sports" of cock fighting and tractor pulls—and the direct descendants of demolition derbies. The audience experiences both jolts of slashing, banging violence and the hilarity of absurd scrap heap machines doing each other in. Robot toys have been around for decades, but they were weak, simple, and did no real damage. Robowar fighters are genuinely dangerous.

We have become used to connecting with events like these, through adroit identification with technology. Since the 1950s children could buy robot toys, which steadily got better. *Sojourner's* 100-meter voyage on Mars in the 1990s, which took an agonizing month to accomplish, enraptured millions.

The adventures of later plucky Mars rovers *(Spirit, Opportunity)* took them through many-kilometer journeys lasting years. With telepresence, human guidance, even time-delayed by about half an hour at Mars, is developing very quickly. The operators command destinations and software takes over the routine navigation and piloting.

The Gulf War of 1991 and then the Second Gulf War both provided robo-conflict without Allied blood. In the First Gulf War the machines died (at least on the Allied side) far more scenically (smart bombs, and so on) than the few Allied casualties; Iraqi losses got much less play. In the Second Gulf War several hundred robots dug up roadside bombs, and a robot attack plane, the Predator, quietly prowled the skies day and night, inflicting casualties, usually without warning. In 2005 the first robot bomb disposer appeared in Baghdad; it could shoot back with good aim at one thousand rounds a minute. A soldier nearby controlled it with a wire-less laptop, the first offensive robot used in combat.

Some got destroyed. Newscasters then used "kill" to describe the destruction of both people and machines. Later in 2005 came robot infantry, able to assault and fire while moving forward on tanklike treads. This got very little media coverage, but plainly, more automation is to come.

Commanding these at a distance recalls a video game quality— indeed, the troops using them have a long background in such skills. Fresh into the millennium, this trend gave us TV's *Battle-Bots,* with its three-minute slam-bang bouts.

Here is violence both real and absurd, calling up memories of the delicious humor of the old Road Runner cartoons. Robots as-sault each other with clippers, buzz saws, spikes, crushing jaws, and other ingenious devices, often cobbled together from domestic ma-chines like lawn mowers and power tools. In their BattleBox they can use any strategy, inflicting mortal wounds, while the human audience sits safely beyond the BattleBox's shatterproof glass walls, watching flying debris clatter against the barrier. The Box has its own tricks, with sledges, rods, saws, and other bedevilments, which pop up randomly to wound one or both of the combatants.

All this gets played for ironic laughter, with over-the-top com-mentary from the sidelines. The rules of human combat get satirized into weight categories: lightweights below 87 pounds, up to super-heavyweight between 316 and 488 pounds. Their creators range from aging engineers to twelve-year-old junior high amateurs. A certain cachet attaches to one who has made his or her robot from the least expensive parts, and especially from scrap.

More such shows are coming, like the Sci Fi Channel's *Ro-bodeath.* Though the robots now contending have budgets of only a few thousand dollars, inevitably under the pressure of ratings the sums will rise. Roller robots built from blenders will give way to walking, stalking specialists with specifically designed pincers or scythes or guns.

History's arms race between human armies will be rerun in madcap, technofreak fashion on full fast-forward. The warriors will get heavier, their armor thicker. Instead of being run by their creators from the sidelines, there will come competitions for robots that can direct themselves, concoct strategies in real time, assess opponents' weaknesses, and find new uses for their own armaments.

All in good fun, of course. But as the human battlefield begins to accept more machines with greater capabilities, the comic mayhem of BattleBots competitions will blend in the evolutionary chain with the coming of combat robots fighting among humans—and finally, inevitably, against them.

In the 1984 film *The Terminator,* a woman confronted with the nonstop violence between a robot killing machine and a man sent to save her, asks, "It will never be over, will it?" Once machines can fight on equal terms with humans, what social force could stop their use? Worse, if directed by artificial intelligences, would fighter robots not carry out the competition between these two intelligent "species" inhabiting the Earth?

The relentless energy of the Terminator class of robot (Arnold Schwarzenegger) confirms this woman's wary prediction as it pursues the two humans with single-minded ferocity, until crushed by a foundry press. That advanced robotic intelligence could have the fanatical concentration of humans, with immense strength and endurance added, makes their use as soldiers seem inevitable.

Robot Armies?

Most robot research funding comes from the U.S. Department of Defense. Obviously armies would rather lose a machine than a man. Robots don't get hungry, feel fear, forget orders, or care if the robot next to them gets killed. Even better, for the accountants, they have no downstream medical or retirement plans. In 2005 the Pentagon owed its soldiers, sailors, and air officers $653 billion in future retirement benefits, which it had no clear plan to pay. Indeed, each fighting soldier costs $4 million over a median lifetime. Robot fighters will certainly cost less than a tenth of that. They can even be retrofitted later for domestic jobs and sold off.

The Bosnian conflict of the late 1990s was the first campaign fought without a single casualty on one side, because the United States used only aircraft. None were robots, but that lossless victory whetted appetites and has probably set the mold. In 2000 Congress told the armed services to develop within a decade robotic ground vehicles and deep-strike aircraft. The goal is to make about a third of all such machines independent. The goal is combat without casualties.

And some things machines can always do better than people, anyway. Some innovations are already about to be deployed in the field. Many ape animal methods.

Crawlers

The Gecko wall-climber described earlier has a relative—the Micro Unattended Mobility System (MUMS) device, currently under development. It is insect-small, an autonomous vehicle no larger than three inches across and twelve inches long. Robust enough to travel on its own, it can survive high accelerations and decelerations, suffering peak impacts of 1,500 times gravitational acceleration. (In the long run, it might be sent forth from a grenade launcher.)

It even has a tail. Two side-by-side wheels drag behind them an active tail that doubles as an antenna. Its central body houses electronics and a suite of navigation and surveillance sensors, including a modular GPS antenna, communications antenna, seismic sensor, microphone, electromagnetic detectors, and perhaps chemical sniffers more sensitive than a human nose. All this in a package that can escape notice.

The rover's embedded intelligence system will be controlled by iRobotics' own software, featuring, as a brochure has it, "redundant sensing and flexible system architecture." Overlapping and redundant sensing makes systems robust in the face of sensor noise, failure, or unexpected conditions. The rover can observe its own lower performance and notice problems. Self-moving robots often repeatedly run into the same obstacle or get caught in a cyclic path. If caught, this rover, introduces random action to add "creativity."

Again, such mobile sensor systems will first be used for covert surveillance and reconnaissance, but the need to travel unnoticed into hostile environments is not unique to the military. Since MUMS robots do not require airdrop, they can also help out law enforcement that needs to covertly use sensors to collect intelligence during standoff situations.

The next generation will feature combined wearable computers and mobile robots. For military use, the robot becomes part of a reconnaissance team, able to respond to verbal orders with local initiative and intelligence. The robot moves in advance of its human team members, while sending back video images and gathered intelligence.

A soldier will direct and monitor the robot's progress through a wearable intuitive interface, at a distance of about a kilometer. The system will use natural voice recognition, a head-mounted display and head tracking, so the robot will know that the command is, "Go in the direction I'm looking." The soldier will use a head-mounted display with "heads-up" computer-generated graphic overlays. At first they will look like deadly toy trucks on treads, with camera snouts pointing front, side, and rear, a machine gun that can be slaved to the cameras, and the ability to hear and smell. Weighing around one hundred pounds, they will cruise at about walking speed and keep it up for four hours on lithium-ion batteries.

Soldiers will be able to hear what the robot does and maneuver it with a handheld joystick, so combat will ape home computer games. This is no accident. A generation has trained using these entertainments, which in turn have been shaped by market forces to be the easiest and most responsive to use.

Beyond that era, robo-fighters will need less supervision. They will increasingly react, see, and think like people, while going places we could not.

Underwater Rovers

This class of autonomous robots seeks to equal the efficiency, acceleration, and maneuverability of fish. Biologically inspired, they use flexible, wiggling, actuated hulls able to produce the large accelerations needed for fishlike bursts of speed and sudden swerves.

They mechanically approximate a fish's fluid swimming motion and navigate environments previously considered inaccessible.

The prototype, named Dart, developed by iRobotics Corporation in cooperation with MIT's Department of Ocean Engineering, is roughly three feet long. It consists of a series of lined actuators, a spring-wound exoskeleton, flexible Lycra skin, and a rigid caudal fin. Modeled after a pike, its flow-foil mechanism "flaps" to create vortices that produce jets to propel it efficiently.

A microprocessor housed inside the head provides the interface between control electronics and the Dart's body. The software, designed to allow rapid development of embedded routines, lets the driver dictate all swimming, starting, and turning parameters from an off-board computer via a graphical interface.

These swimmers can covertly gather intelligence close to shore. Their fishlike locomotion will reduce power requirements, make detection more difficult, and facilitate escape. On radar and sonar they will look very much like ordinary fish, particularly after "stealth" surfaces appear to outsmart reflected acoustic and electromagnetic waves.

For commercial and research use, where negotiation of hostile environments is essential, they can navigate intricate structures. For harbor cleaning, this will help find hazardous materials. In mining, swimming prospectors will prowl far larger areas than human crews in submarines can. Exploring the deep ocean will open to tough, independent robo-swimmers that can monitor for long times the countless valleys, caverns, and geothermal vents we have only begun to fathom.

Fetchers—Countermine Intelligence

There will soon be a new approach to a global plague—the land mines left behind in wars. A team of low-cost, robotic mine hunters can provide rapid and complete coverage of a minefield. A swarm of robots will ultimately be capable of cooperatively clearing a field of land mines under the supervision of a single operator.

Designed for low-cost duplication, because they can make mistakes and trigger the mines, these robots are just a few years from deployment. Already they have successfully detected, retrieved, and safely deposited munitions in the real world, visiting areas replete with unfavorable obstacles, terrain slopes, and poor traction.

Common problems arise whenever robot teams do a job. How can a lightly trained technician run such a complex system? How can the robots cooperate effectively? Today's IS Robotics' Fetch II robots perform their tasks autonomously, but with a single operator. Learning, "behavior-based" software keeps track of what the robots are doing and anticipates problems. The human notices only when something goes wrong. Software mediates robot–robot interference within the swarm and supports cooperation among them.

Terminator?

The above military 'bots snoop more than they fight. It does not take much imagination to see that modern tanks, outfitted with omnidirectional sensors in many frequencies, assisted by smart software and fast chips, could make their way through a future battleground without humans aboard. Current Pentagon plans are for combat units to have robot complements making up less than 10 percent of the "troop" strength.

How good can they become? Much science fiction features fighting machines of the future outwitting human antagonists, and even cyborged people with formidable abilities of their own.

Perhaps this could happen, as munitions become smarter and warfare more mechanized. Just imagine what power source could run a Schwarzenegger-size machine that can fight for even the duration of a two-hour film. Lithium-ion batteries won't do it.

Experts like Robert Finkelstein, president of Robotic Technology, have told the Pentagon that a true robot that moves, thinks, and fights like a soldier will not appear on battlefields for another thirty years. Today's best attempt is a boxy prototype on treads, with a cyclops eye. Its right arm is a gun and its left is an all-purpose tool that can open doors, lift blocks, and cut holes. Told to fire, it locates, identifies, and then quickly shoots. It can hit a Pepsi can ten meters away.

Of course, it will be quite a while before a shooter robot gets an order to find, identify, and kill a human enemy all on its own. Humans make the trigger-pulling decision.

For now, we get good performance by specializing machines. We have scouts that can prowl buildings, caves, and tunnels. Others drone overhead, staying patiently aloft for tens of hours. Big haulers carry weapons and gear, while others scout and report back. Others will endlessly follow their rounds on security watch, often in the dark since they can see by infrared and we can't. More savvy types will sneak behind enemy lines, eavesdrop, even conduct psychological war—making the enemy always look over his shoulder at every odd sound or movement wears him down.

But the need for an all-purpose machine will persist. Robot intelligence is increasing, as chips shrink and software gets smarter. Perception is the fulcrum of improvement. With a bit more progress, quarter-ton trucks will have robot drivers in combat zones. With digital road maps and global positioning satellites, robot convoys are only a decade away.

Today's robots work at the level of perception of not terribly bright mammals. In a generation, robots will work at the level of primates. At that level, it will be possible to let machines fight on their own. *Monkey see, monkey shoot.*

Still, many conflicts are messy matters of mud and blood. Machines cannot easily fight in trenches, snow, jungles or in house-to-house, hand-to-hand guerrilla conflicts. Robots will not fare well there, amid grit, smoke, and rust.

So pressure will make them better—more rugged, savvy, perceptive. Doctrine always lags behind technology—the longbow, cannon, tank, plane, and nuclear weapon all outran the strategies first used to employ them. So it will be with machines. Asimov's Three Laws will not apply to a combat robot, so they will need no tricky moral calculus. But they will need to tell friend from enemy, a surrendering foe from a fighting one, and enemies lying doggo. Our doctrines will change, too: Will few casualties on one side make war with technologically inferior societies more tempting?

And what of the robots? If the machines are smart enough to outwit humans amid difficult terrain, they might very well have to be smart enough to question why they are doing it—a point seldom noted by filmmakers, who assume all advanced machines will still be absolutely obedient and have no desires other than perhaps malformed human motivations.

A common movie idea, which applies so broadly it includes many tales of alien contact, is the Menace Theme: *An intelligence we do not understand goes crazy (by our definitions, but maybe not its own). So it does evil things outside our moral code—mostly destruction of people and cities. So we have to kill it, and then the tale is over.*

Robots are just one category of menace. Why do we like this idea so much that it has spawned hundreds of films?

Perhaps it's because we derive some unacknowledged gratification from watching the destruction. Many love Godzilla, even though he has a grudge against Tokyo. We watch *BattleBots* not out of love of robots, but of the smashups. We can wash our hands of any guilty feelings because they are just machines, after all. Though they might get smart, they won't be human.

This extends to warfare. Robots can take the risks for us only in stylized, well-defined physical situations. The advanced nations will probably seize upon this in future, trying to make their conflicts resemble the Gulf Wars rather than Vietnam. Their antagonists will do the opposite, trying to pin down vulnerable infantry. The success of the NATO air war against the Serbs in 1998 shows that even messy conflicts can be won with high tech, especially if one attacks the obvious, fixed economic infrastructure rather than only troops in the field.

Robots will make these contrasts ever greater. We will always see men with guns and bombs seeking power, but as the technological gap between societies widens further, such groups will have to resort to terrorism (which itself gets ever more complex and technological) to make their bloody points. Against them will stand robots of ever-greater sophistication, patience, savvy, and strength. Under enemy fire they will haul ammo, reconnoiter, search buildings, find the wounded.

They will have many shapes—crawlers like caterpillars or

cockroaches, heavy assault craft like tanks or tractors, fliers looking like hummingbirds, or even "smart dust" swarms of robo-insects. Some will resemble animals and insects, to escape notice. Others will intentionally look bizarre, to frighten or intimidate. Few will be able to pass as human, even at a distance and at night, for quite a while.

Their inner minds will be odd, stylized, but steadily improving. We may come to see these metallic sentinels as our unique heroes, the modern centurions. The other side will see them as pure walking terrors, killing whatever romance might still be left in war.

Or perhaps not. For we do have some historical precedent to instruct us. Medieval warfare in the centuries-long age of knights developed conventions quite unlike those we know in too many modern wars. Knights required a large support team, a hundred or more who carried out the heavy-lifting jobs in the logistics of horse and armor. These were in the army, but were kept outside the bounds of battle, and even if overrun were not killed—though their gear might get stolen. Knights themselves were fair game, but here, too, a thrifty ethics ruled. They were most often not killed but instead cornered or injured, then captured and ransomed for large sums; then they could fight again, for capture was no disgrace.

The prevailing rule was, *fight only the fighters.*

No one attacked the camps supporting the knights, or executed prisoners, since they could be ransomed or sold as slaves. To kill noncombatants was an atrocity, often punished. So until around 1650, European war was a conflict of big metal war machines that happened to have humans inside.

This suggests a strategy: Remove the humans, use robots in combat wherever possible, and knowingly drive the war culture toward a different moral standard. Use international standards, such as the rather outmoded Geneva conventions, to create a new view. We could see the eventual evolution of robot warfare back to such a code. Of course, medieval times had plagues and starvation that ran alongside wars, but these messy side effects exist now, too; the Four Horsemen of the Apocalypse often ride together, led by War. A semi-medieval code would be in some ways superior to our current style of total war. The second half of the twentieth century saw common terror, atrocity, and wholesale destruction, even in "advanced" nations like those in the Bosnian-Serbian conflict, which lasted a decade and resulted in half a million slaughtered.

A robot war culture does not have to be worse than our moral standards today. That may seem a radical conclusion, given the pervasive imagery of *The Terminator, RoboCop,* and similar films. But it is important to believe that our future can be better than the worst-case scenario. Indeed, it is essential, or else, why bother?

CHAPTER 7

The Rights and Wrongs of Robots

When are they no longer just appliances?
And what will they think of us?

In advanced societies, war is a distant rumble at most. Robots who work around us in hospitals, warehouses, and even our backyards will have far more impact on our lives in the next few decades.

Quite quickly, some will attribute personality to these, giving them names, expecting them to act somewhat like us—and being surprised when they do not. This is likely to be the source of much reflection, confusion, and comedy. Within a decade, we can expect to see a situation comedy with a robot character, played by a robot. It may resemble the 1970s *Mork & Mindy,* though without Robin Williams's manic comedy. Probably the robot will be a straight man, with the occasional startling insight into the foibles of humans.

Why do we so readily identify with objects we know to be programmed machines?

Personhood

Dr. Anne Foerst is a theologian, a Lutheran minister, who works in the Artificial Intelligence Laboratory at the Massachusetts Institute of Technology. As director of MIT's God and Computers project, she is the theological adviser to scientists building robots intended to interact socially with humans.

She has considered the ongoing debate over whether computers can really think the way we do, and stepped aside. Her interests lie not in scoring points in philosophical disputes but in our experienced realities—the reactions that tell us what we really think. Her work anchors in our perceptions of existing robots, and how we ascribe traits to them.

INTERVIEW WITH DR. ANNE FOERST

Q : What does it mean to be a person? Can a machine *be* a person, not just resemble one?

A : When you look at critiques against AI and against the creation of humanoid machines, one of the things which always comes up is "they lack soul." That's the more religious terminology. The more secular terminology is "they lack consciousness."

Ultimately those terms, at least in those eyes of the critics, are the same. They seem to say that humans have something special, that we are distinct from all the other animals, and distinct from machines. Therefore, we can't rebuild those things by mechanical means alone.

I asked myself, where does this wish to be special come from? Because when we look in the Hebrew Bible, we find that soul, or *nefesh*, is not something you *have*, but something you *participate in*. In a healthy community you will be with fellow humans and God. So "person" is not defined to mean having a soul, but to participate in the community.

Q : Defining by process.

A : That means "person" is an *assignment,* given to each one of us by our parents and our closest community right after birth. It is given to us by God in the first place, and we are free to assign it to others. But we are also free to deny it to others.

Q: Can we assign it to a machine? Can a machine be a person in this sense?

A: I think a machine can definitely be a person. The more social our machines get, the more interactive they are, the more they learn out of interaction with us, the creators, and the more we interact with them. For me there is no question that at some point they will be persons like anyone else. They will not be *humans* because they are different material, but they will be part of our community, there is no question in my mind.

Q: There is no division between us and them?

A: To be human is a biological category, because of the genes we have. Robots will not have that, even though they might be in part based on biological material. But will they be part of our community, yes—because the communities we live in are basically very, very small. Each of us only assigns personhood to a very few people. The ethical stance is always that we have to assign personhood to everyone, but in reality we don't. We don't care about a million people dying in China of an earthquake, ultimately, in an emotional way. We try to, but we can't really, because we don't share the same physical space. It might be much more important for us if our dog is sick. So personhood depends very strongly on sharing physical space, interacting with each other, creating stories together.

Q: Since we have denied people personhood, is it wrong to deny a robot personhood? The MIT robot developed by Cynthia Breazeal, Kismet, apes human facial moods with uncanny ability. Maybe this comes from our human ability to extract lots of meaning from subtle clues, as when we look at clouds overhead and see objects, or the "face" of the man in the moon. But is that enough to establish a "personal" relationship—or are we just responding to a copied cue?

A: To think about robots and personhood we have to at first see what this whole concept of personhood is about. When I look at the human system and its functions from the cognitive side and from anthropology, I have two different angles:

One angle is really how our biological system works. There are strong limitations to how many people we can have a personal relationship with, approximately 150. In a world where we are constantly surrounded by hundreds of thousands, we have developed quite a few mechanisms to avoid seeing personhood in people. By not recognizing their faces, by not remembering their names, but just putting them into categories.

So when we interact with the cute robot Kismet we are creating stories together. Then we can't help but start a personal relationship with that robot. It is just amazing to me. I have interacted with Kismet so often, but each time I kind of fall in love with it again. Even though it can't yet learn and it is still fairly primitive, imagine that those creatures will get more and more autonomous and able to learn. They will recognize you. You really get emotionally involved with them.

So why not assign them personhood? Usually the argument is, "Well, we are much more than machines." What does that mean?

We are biological systems, built up from certain material, we function in a certain way. What makes us really human is that this body here is socially connected, embedded in the world and in the social community. That's what counts. To enable the robot to have the same kind of embeddedness and social interaction would make them too personal for a lot of people.

Q: I see—they would not like to have robots doing a range of personal things for them? Taking out the garbage, maybe, but not helping host parties, keeping track of our medications, or making our dinner engagements? Nothing too personal? So can a machine be said to have a psyche—a soul?

A: A machine can have a soul if we define soul not as this little something that makes *us* special, this irreducible essence. If we interpret soul in the biblical sense, especially in the Hebrew Scriptures, it is not independent from the body. Soul emerges out of social interactions between individuals, their group (the Jews) and God. Soul is saying, "I assign you personhood and you assign me personhood because we both have been assigned personhood by God. We have been given by God intrinsic value." It is a gift, and therefore we are free to give it to others. If robots are part of that interactive community, then I think we have to assign them the same thing, too. This is not to say that robots will be clearly seen as persons in any objective fashion, or by everyone. No. Because personhood is not an empirical fact, something we know every human has—I mean look at the phenomenon of racism. Just because someone has a different skin color from mine, I may not see them as persons. I see them as *humans* but not as *persons.*

Q : Can we see a robot as alive? Suppose one day a robot says suddenly, "I'm conscious, you know." We take a person's word for it; how about a robot's?

A : If it behaves alive, it is.

Q : But we know about the interior states of humans because they're like us—we think.

A : Even though our mind might know that this is just a machine, our behavior towards another creature is very intuitive, developed by evolution over hundreds of thousands of years. And there were no robots about. Now we will have to deal with our own creatures, who behave as if they were alive, social, loving, and emotional. I think we aren't even emotionally capable of distinguishing between alive, animated, or stuffed. In a way it doesn't really matter. What does matter is the question, "Will they ever be so complex that some of us will treat them as persons?"

Will there ever be the point where we cannot switch them off anymore? This is not an objective answer. I think if the engineers, who know exactly what is going on, at some point decide that something is in there, we cannot really switch it off anymore. That will be certainly a big criterion for them passing a threshold.

The difficulty with all definitions of personhood that come from social standing is: What if that gets removed? This can happen accidentally, if a Robinson Crusoe figure gets isolated for years. Hermits do so deliberately—but do they give up their right to demand personhood from us?

The classic way we turn our fellows into the Other is through withdrawing their humanity. In-group stigmatizes out-group, preparing the way psychologically for conflict, warfare, or even genocide. Chimpanzees seem to follow a similar course in their definitions of their groups. If robots come under social pressure, perhaps there will be calls for defining them, no matter what their apparent intelligence might seem to be, as nonpeople, having no rights.

The social definition seems open and accepting, but it can just as easily turn into rejection. Defining so intimate a quality by external connections carries risk. "It is a gift, and therefore we are free to give it to others." The unspoken corollary is, *And withhold it, too.*

Americans especially view personhood as an internal quality, not an external one. That is why the U.S. Constitution takes the position that rights are inherent, precisely so that they cannot be removed by some future social consensus. This turns to the issue of defining human, as usual, with a special slant: Can "persons" mean something other than human?

At what level, then, should we assign personhood? (After all, in law, corporations have such status.) We have already met this problem in real life, and have our rough-and-ready rules. Does a bacterium have a mind? No. Insects? Still no, though they display social behavior. Dogs? Weeeelll . . . they seem to have feelings, and so some level of mind. The comatose? There are some brain waves, so there remains a potential for a mind to recover from its injury.

Beyond that, we have no experience with machines that could inspire confidence. The coming proliferation of robots that mimic human moves, expressions, emotions, and responses will confuse the issue greatly in our commonsensical world.

What may matter more than abstract arguments is the machines' physical presence. We have an innate bias for things that move, respond, have their own bodies. The persona of HAL in *2001* could act in the real world by controlling machines, but it had no embodiment itself. The audience of the film did not seem to mind HAL's extinction, even with it singing "Daisy, Daisy . . ." as it died.

Machines that move may be harder to dismiss. If they interact with our world via sensors, processors, arms, and heads, they will seem more substantial, more nearly "natural" even if they obviously are made of machined parts. The chips running them may even be inferior in capacity and range to those that run, say, a sophisticated mathematical program. Indeed, we now have software (Mathematica, MathLab) that performs miracles of computation that would have astounded the great calculator mathematicians of the eighteenth and nineteenth centuries, such as Gauss. We would not think of extending the recognition of personification toward Mathematica, however brilliantly it solves integral equations for us in a few seconds, and even graphs the results in three colors, saving days of labor. Obviously it is just a tool, made to do one thing only.

Instead, we favor the concrete. Computer programs cannot compete for our attention with even a simple toy robo-dog, bought for less than one hundred dollars. Perhaps evolution has selected for this, anchoring us in the physical world, not the abstract.

But what of a robot that could compose and (digitally) perform lyrical, flowing music that captured the ear and lifted the heart? Suppose it was just a fixed box of chips, no human face or legs to draw out our empathy. Is this intangible intellectual world less compelling?

This thought experiment poses a new form of the mind versus matter problem that has vexed philosophers for millennia, one likely to arise if computational intelligence ever becomes real enough to rival a Beethoven. Creativity seems to many a clear case warranting the award of personhood, no matter what the origin—a compelling standard, because we see it as our highest function, beyond mere logic.

Perhaps human nurture can elicit from machine and computer nature ways to create unique companions. From that may emerge fresh creativity. We humans have longed for companions, calling them pets and imagining angels and aliens. But maybe we can create our own, or discover higher functions in animals we think we know well.

Indeed, elephants can produce both paintings and music that seem coherent and esthetically interesting. This fact points to some degree of consciousness that may warrant our award of rights. Does art imply consciousness?

This is a social issue that may provide some guidance to us in the decade ahead. Which will attain rights first—machines, or chimps and elephants and dolphins?

Should There Be Robot Laws?

The first attempts to think constructively about how to deal with humanoid forms that were quite different from us—including cyborgs, androids, and robots—came from science fiction. Robots present the most extreme case, with no fleshy components, so they attracted the vivid imaginations of such early thinkers as Isaac Asimov. Used to thinking systematically because he was a trained biochemist with a Ph.D. from Columbia University, over several decades Asimov wrote a groundbreaking series of stories built around his fundamental Three Laws of Robotics:

1. A robot may not injure a human being or, through inaction, allow a human being to come to harm.
2. A robot must obey orders given it by human beings except where such orders would conflict with the First Law.
3. A robot must protect its own existence as long as such protection does not conflict with the First or Second Law.

These Laws shaped much thinking, both in science fiction and in robotics, for decades. Only now, over half a century since they were worked out in an ingenious series of stories that tested each phrase of the Laws, can we see them as projections of the anxieties and assumptions of their time. Rather than universal Laws, they are really rules for behavior. They center around several implicit attitudes, ones that can affect any partially artificial being (androids or cyborgs).

The message of these laws is that avoiding evil robot impulses is crucial, as though they would naturally arise among any thinking entities. That assumption glares out from the 1940s, when certainly recent history seemed to give ample proof of seemingly inherent human evil. What's more, the animal kingdom—"nature red in tooth and claw," as the poet has it—echoes with the cries of those who came in second place in the struggle to survive.

Asimov noted that as seemingly basic an instinct as self-preservation would need to be introduced into a mind that one was building from scratch. Nature gives animals the savvy to stay alive, but machines must have this inserted. Generally they don't; in the decades since the Three Laws were worked out, we have built "smart bombs" and cruise missiles that happily commit suicide while damaging our enemies. This gets patched up in the Third Law, presciently.

The Second Law is actually needed to enforce the First Law—otherwise, how would a robot know that it must obey? Robots ordering other robots must not override human commands. (But with the advent of cyborgs, how is a robot to know a true human? That is an interesting channel for future stories. One can imagine a cyborg demoted to less-than-human status because robots refuse to recognize him or her.)

The Three Laws of Robotics are in fact moral principles disguised as instructions. Compare them to the Ten Commandments, which are much more specific: *Honor thy father and mother. Thou shalt not kill.* Of course working out how to use these (when is it okay to kill in warfare? and who?) demands more interpretation.

But then, so do the Three Laws. In fact, rather than guides to how to build robots and program them, they are better seen as what they originally were: a neat way to frame a continuing series of stories, each testing the boundaries of the Laws. Real robots need much more specific engines to tell them how to work. How to do this is still unknown. Humans know the law and obey it if they choose. Robots we want to obey the law always—with superior strength, endurance, and ruggedness, they could be terribly dangerous, as Asimov feared.

But we do not know how to force such compliance in a machine that still has to have some measure of autonomy. Perhaps we never will, because there could be an inherent tension between independence of mind and law abiding.

To show originality, a computer must be able to get outside its normal routes of operation. Novelty demands that the rule set must not be predictable. This means that the outcome will be unreliable in the sense of unpredictable, new. So probability of error grows as innovation increases.

Intelligence defined this way may very well imply fallibility—just the opposite of what we usually expect of our machines. Indeed, it seems plausible that such a machine would also not always be able to make up its mind. In this regard, too, robots would be much like us. We may have to accept some danger as a trade-off for some degree of robot autonomy.

Our Complicated Selves

We now realize, after nearly half a century of work, that the tough problem is how to instill motivations in other minds. Getting robots to obey our laws may fundamentally emerge from getting them to do anything at all.

Survival is not one task but a suite of ever-alert programs that have to interact with the ever-shifting environment. So are other, milder motivations like empathy and cooperation.

This realization, that our commonplace urges are really quite complex, has made us see that many supposedly simple human tasks are very complicated. Take, say, picking up a cup of tea and then not sipping from it, but instead blowing across the top because we can tell it is too hot. This is in fact a feat of agility, sensing, and judgment that no machine can presently do nearly so well. (Nor does a machine store the memory of burnt lips from a decade ago, which pops up as a warning when we reach for a cup.) Indeed, if one computer can do all that, it must be specifically designed for that job and can do nothing else. Yet we can sip tea and read the newspaper, half listen to our mate's breakfast conversation, and keep breathing.

Such intricacy is built in at the foundations of our minds and bodies. Life is tough; we must do several things at once—or else more versatile creatures will do us in.

But do we perform such adroit tasks as (to echo an old joke) simultaneously walking and chewing gum, all by following rules? This is the contrast between *knowing how* and *knowing that,* as Keith Devlin of Stanford University puts it. The fundamentally Cartesian emphasis on following rules to order our actions implies a central question: *Can* we work that way? *Do* we?

It seems logical; just follow the directions. But in 1670 Blaise Pascal, the mathematical philosopher, saw the flaw: "Mathematicians wish to treat matters of perception mathematically, and make themselves ridiculous . . . the mind . . . does it tacitly, naturally, and without technical rules."

But we do not ride a bicycle by following serial rules; we parallel many inputs and respond in ways not yet understood. Here is where the entire agenda of rules-run intelligence runs into a deep problem.

If tasks are done sequentially, they run the risk of not getting done fast enough—so the only answer is to speed up the computer. This acceleration can lead to huge problems, because neither humans nor computers can simply step up their speed to meet every problem. Our brains have engineered parallel processing (solving problems by running separate programs simultaneously) to keep up with the real world's speeds. They can't just add new lobes for new problems, except on an evolutionary time scale of millions of years.

This also relates to the top-down, rules-from-above approach to artificial intelligence. Perhaps that approach is fundamentally limited because a rules-based mind would be too hobbled and slow. The alternative, starting with systems that learn in small ways and build up a concept of the world from direct experience (as in mobile robots), may work better. Nobody knows as yet.

One can imagine a robot brought to trial for some misdeed, perhaps injuring a human. So far we just reprogram or turn off a defective machine. The legal status conferred by a trial would be a watershed—an implicit concession of rights, in fact.

Since an autonomous robot is allowed freedom of movement, it is held liable for its acts. Such a trial would then define robots as of human status, held accountable to human law. Persons, then—who will have to act that way.

Will Robots Feel?

The robots we meet in the next few decades will not look like mechanical men, the classic science-fictional image. There were good storytelling reasons to make robots humanoid, to get the audience to identify with them. Some current robot builders betray the same need to make their machines look or move like humans (as with the MIT facial robot, Kismet, and its successor, Leonardo). We have experience in dealing with humanlike others, after all. But perhaps the essence of robots will be that they are not like us, and we should not think of them that way, however appealing that temptation might be.

We are trained by life and by society to assume a great deal about others, without evidence. Bluntly put, nobody knows for sure that anyone else has emotions. There are about six billion electrochemical systems walking around this planet, each apparently sensing an operatic mix of feelings, sensations, dazzling delights—but we only infer this, since we directly experience only our own.

Society would be impossible to run without our assumption that other people share our inner mental states, of which emotions are the most powerful. Without assuming that, we could anticipate very little of what others might do.

Robots bring this question to the foreground, even more than cyborgs. How could we tell what a robot would do? Of course, we could install Asimov's Three Laws, and pile on maybe dozens more—but working out what will happen next is like treating life as an elaborate exercise with an instruction manual. Nobody thinks that way, and robots would be rule-bound catatonics if they had to function like that.

Should we want a robot to take up the task of acting so that we could pretty reliably predict how it would feel? That ability is available already, without the bother of soldering components together in a factory—it takes only two people and nine months, plus a decade or two of socialization. Surely we do not want robots to just act like us, remorselessly, if they are to be anything more than simple slaves.

Suppose, as good Darwinians, we define emotions as electrical signals that are apt to make us repeat certain behaviors, because those increase our chances of reproducing ourselves. Then feelings look almost like computerized instructions, overriding commands—and machines can readily experience these. Suppose that in a decade your personal DeskSec comes built into your office, and from the first day quickly learns your preferences in background music, favorite phone numbers, office humidity, processing typeface on your computer monitor—the works. A great simplifier.

Then this Girl Friday might get quite irked when things don't go right, letting you know by tightening "her" voice, talking faster, maybe even fidgeting with the office hardware. Does the DeskSec have emotions as we understand them? To answer that, we must assess how realistically the DeskSec does its acting job.

That is, is it acting? All such questions arise from the tension between internal definitions of mental states, and external clues to them. We imagine that other people feel joy or pain because they express it in ways that echo our expressions. Of course, we have a lot of help. We know that we share with other humans a lot of common experience, from the anxieties and joys of growing up, to the simple pain of stubbing a toe. So with plenty of clues, we can confidently believe we understand others. This is the task met by good actors—how to render those signs, verbal and physical, that tell an audience: "See, I feel this, too."

Even so, juries notoriously cannot tell when witnesses are lying. We can't use our sense of connection to others to get reliable information about them, because people know how to fake signals. Evidence accumulates that even our nearest surviving relations, the chimpanzees, do not readily ascribe to their fellows (or to us!) an inner consciousness. We would say that they have little social awareness, beyond the easy signals of dominance hierarchy.

Chimps *can* build a model including human awareness, though.

This ability is explored in experiments with them, which test whether they really do register where human attention is directed. They quickly work this out, if there is a reward in store. (This skill may come from trying to guess what the alpha male of the tribe is going to do next.) But they do not naturally carry this sense into their ordinary lives. Taught to realize that humans facing away from them can be looking over a shoulder, they respond to this fact to get bananas. But a year later they have forgotten this trick; it's not part of their consciousness tool kit. Human toddlers can think that others see what they see, so if they cover their eyes, you can't see them—but once they catch on, they never forget. Chimps do.

These are the sorts of tests we should apply to any robots who petition to be regarded as human. Such tests are rigorous; perhaps only the dolphins can pass them now.

When and if robots can compose symphonies, then we'll be on the verge of asking serious questions about the inner experiences of machines. If we decide that robots have a supple model of us, we may have to ascribe humanlike selfhood to them. Aside from legal implications, this means we will inevitably be led to accept, in machines, emotions as well as abstractions.

Not that this will be an unalloyed plus. Who wants robots who get short-tempered, or fall in love with us?

Inevitably, robots that mimic emotions will elicit from us the urge that we treat them as humans. But we should use the word "mimic" because that is all we will ever know of their true internal states. We could even build robots who behave like electronic Zen masters, rendering services with an acute sense of our human condition, and a desire to lessen our anguish. But we will not know that they are spiritual machines.

Probably someone will strive to perfect just such robots. After all, why should robotic emotions not be the very best we can muster, instead of, say, our temper tantrums and envy?

In chapter 3 we argued that emotions are a vital part of our psyche. We have no idea how an intelligent mind of any subtlety would work without emotions. Humans with disabled emotional centers do things that seem rational to them, but in their lack of foresight and insight into people seem absurd or even suicidal to the rest of us. That is the implicit threat many feel about intelligent, emotionless robots—that they would be beyond our understanding, and so eventually beyond our control.

We may be forced, then, to include some emotional superstructure in any advanced robot "psyche." Perhaps the inevitable answer to *Will robots feel?* is "They'll have to—we'll demand it."

Will Robots Sleep?

All along, philosophers and computer mathematicians have told us that our uniquely human skill at juggling symbols, particularly words and numbers, defines us. Small surprise that they happen to be good at this themselves, and in believing these abilities define the pinnacle of creation, think that they have captured consciousness.

This belief is comforting, and goes back to Plato and Marcus Aurelius, who commanded, "Use animals and other things and objects freely; but behave in a social spirit toward human beings, because they can reason."

But other, simpler definitions can illuminate how robots may behave. Humans are not just symbol-movers. One of our least noticed traits is that we fall unconscious every day for many hours, while many animals do not. Is sleep important?

Living on a planet with a single sun, and a pronounced day–night cycle, has shaped the biology and ecology of almost all animals. One must say "almost" because the deep-sea ecology is uniformly dark, and yet sustains a surprising complexity—witness the thermal vent communities.

As day-living, light-adapted creatures, we are most familiar with the other day inhabitants; but at night, in the ocean as well as on land, a whole new suite of animals emerges. Among them, owls replace hawks, moths replace butterflies, bats fly instead of most birds, and flying night squirrels replace almost identical day-living ones. On coral reefs all manner of creatures emerge from sheltered recesses when night falls.

Animals without backbones, and the slower, cold-blooded vertebrates, do not indulge in sleep as we do. They hide and rest for a few hours, but display little change in neural activity while they do so. This fits with the idea that it is smart to stay out of the way of predators for a while, and that some rest is good for any organism. Still, these periods among the simpler orders of life are brief, a few hours, and carry no mental signatures of diminished brain activity. Quite probably, the defenses are still running, ears pricked for suspicious sounds, nose twitching at the unfamiliar scent.

Even among vertebrates, only mammals and birds have a characteristic shift from fast waves to slow ones in the forebrain, the typical signature of deep sleep. Probably this is due to the great development of the latter groups' cerebral hemispheres. The simpler brains could not display the advanced signs of sleep, because they do not have a cerebral cortex, and do not shift wave rhythms.

Indeed, sleep is risky. Like consciousness, it demands time and body energy. Nature does not allow such investments to persist without payoffs, so both traits must have conferred survival capability far back in antiquity. On the face of it, lying around in a deep torpor, exposed to attack, does not sound like a smart move.

Yet we and other mammals cannot do without our sleep. Deprived of it, we get edgy, then irritable, then have fainting spells, hallucinations, and finally we collapse or even die. Sleep can't be a simple conservation move, either, because we save only about 120 calories during a full eight hours of lying insensate. Even for the warm-bloods, that's not a big gain; it equals the calories in a can of Pepsi.

It's also unlikely that nature enforces true sleep solely to keep us from wandering around in the dark, when we are more vulnerable. There is too much variation in how much sleep creatures need for this to be a reliable precaution.

If the day–night cycle imposed by the planet were the primary reason for an enforced downtime (an ecological reason), it seems likely that evolution would have taken advantage of it for purely biological reasons. For example, plants undergo dark time chemical reactions that ultimately trigger flowering at a precise time of the year. Animals, too, would've "invented" things to do during an imposed rest period.

So which came first—the ecological or the biological reason for sleep? It's a chicken-and-egg kind of argument that science hasn't answered with finality.

In any case, large animals and birds must sleep, even when they have no ready shelter, or prospect of any, as in the African veldt. Horses sleep only three hours a day, with only about twenty minutes lying down, but they would be safer if evolution let them stay awake all the time.

Sea otters, air-breathing mammals living precariously in the ocean waves, tie themselves to giant kelp and sleep half a brain at a time. One hemisphere sleeps while the other literally keeps watchful eye out for danger.

Sleep seems basic. We process memories while dead to the world, throwing out some and storing away many fewer for later use. We arise refreshed, probably because sleep has tidied up and repaired some sort of damage that consciousness does to our brains. Take that processing and neatening-up away, and we work less reliably and get sick more often.

This correspondence between sleep and consciousness suggests that animals slumber because they have some need of repair work, just as we do. Plausibly, the daily waking state of mind among animals that must sleep resembles our mental frame of the world, the modeling we call consciousness. This seems a sensible explanation for our intuition that our mammal pets have some kind of consciousness, interpreting their world in ways we understand automatically—as, say, when a dog tugs on his leash as he nears a favorite running spot, giving all the signs of joy and anticipation.

Since consciousness has evolutionary utility, and sleep cleans up after consciousness has messed up our minds a bit, we must see these as parallel abilities, each making its contribution to our survival.

A natural conclusion, then, is that conscious robots will have to sleep. They will not be tireless workers like the present automatons in car factories, riveting doors to frames around the clock. "Useless" sleep hours must be budgeted into their lives.

The same then holds true for artificial intelligences. Mathematicians have long seen these as complex devices for carrying out programs, called algorithms. But if robots must be refreshed, sleep is probably only one of the necessities. We do not keep people trapped in rooms, laboring incessantly when they are not catching their zzzzzz's. Not only would they protest, but they would get dull, listless, and inefficient, as well.

So robots and even computer minds will probably have to have regular outings, vacations, time off to recreate themselves. This will make them seem far more humanlike to us, of course, because they will be exactly so. As philosopher Matt Cartmill notes, "If we ever succeed in creating an artificial intelligence, it's going to have to be something more than just an algorithm machine." How much more, no one knows as yet. Probably it will be much more like ordinary workers, needing time to laze around, be amused, distracted, and relaxed. Maybe they'll even watch mindless daytime television.

CHAPTER 8

Us and Them

Will humans have to defend themselves against robots?

People seem to especially like to order others around.
That may be the greatest social use of robots.

—Isaac Asimov, in conversation

If you were in the mind of a centipede, would you think like a human or an insect? Could you be human in a body of a wholly different design? Is there such a thing as "body intelligence," that determines how we think and feel, and why does MIT's Artificial Intelligence Lab have a theologian on the staff?

Telling *us* from *them* is not restricted to national politics. Much energy and inventiveness in Nature deals with making this vital distinction at many levels.

At the individual level, the body's immune system was designed to distinguish between "of the body" and "alien." An elaborate system of recognition and defense against foreign antigens is buried deep in the genes throughout the animal kingdom.

Behavioral signals like birdsong or fruit fly courtship dances allow like to recognize like, facilitating mating between members of the same species, and discouraging cross-breeding. Making the wrong signal can have disastrous consequences: cute little rabbits will harass one of their number to death if it seems "different" in the eyes of the group.

Precivilization humans lived in small cohesive groups that later became small villages. In some societies, incessant warfare between adjacent villages was the norm.

Keeping track of relatives has always been important to reduce the possibility of inbreeding. Human societies had incest taboos long before we understood the biological mechanism behind them. This sets up dueling imperatives: you want to live with your group, but you need to marry outside of it. Promoting your relatives' health and well-being helps your shared genes get passed on. But on the other hand, you have to mate with someone with different genes so your own offspring will thrive.

Today, huge metropolitan agglomerations have replaced scattered villages, but the old impulses are still at work. People identify themselves by their neighborhoods, church affiliations, or through shared activities, like hobbies, the PTA, or playing on a sports team. And they use these allegiances to distinguish themselves from others. The bane of modern civilization is a core of ethnic, cultural, and racial intolerance, remnant impulses from our past.

How will these ancient impulses play out as humans confront humanoid robots?

Dr. Anne Foerst, as director of MIT's God and Computers project, is the theological adviser to scientists building robots intended to interact socially with humans.

INTERVIEW WITH DR. ANNE FOERST

Q: Why build machines that are humanoid? We have plenty of humans.

A : In Japan, they have an aging society like all other industrial nations, but they are not going to get new people in because it is a
very closed society. So if they build robots to support the elderly
in their homes, they have to be humanoids because houses are
designed for humanoid shape.

The reason *we* do humanoid machines is slightly different.
Cognitive science tells us that the shape and form of the body is
crucial to our way of thinking. A humanoid body will experience the world in a similar way to us, and therefore will develop
similar ways of thinking and of interacting with the world. A
horse thinks differently from an ant, and an ant thinks differently from us, because we all have different body needs, and different challenges from the environment. In order to think like
us, the bodily form of the robots has to be as similar as possible
to us to experience the same body experiences.

The other reason for humanoid form is that, to become intelligent, we need to be treated as persons right away. This is what
happens between a newborn and its parents. A newborn doesn't
have any motivation, thoughts, self-awareness, or consciousness, only very basic emotions. Yet we treat it right from the start
as if it does. And because we treat it that way, it has a chance to
become all that.

Our reaction to a newborn works because babies look the
way they do—big eyes, round face, big forehead, big head, all
front, and small body. By building all those features, called the
Baby Scheme, into Kismet, we have a robot that we want to assign motivation and consciousness, et cetera. And when Kismet
starts learning, it might actually develop in the same way as a
baby. In short, if you look like a human, you are treated like a human, which is crucial for the development of human intelligence.

There seems to be widespread agreement on this last concept.
It's worked out in the Lawrence Durrell novel *Nunquam* (1970), a
story of the creation of an artificial woman modeled after a dead actress and invested with her memories.

The bed she lay in was a long white surgeon's operating table with gleaming leverage members in tubular steel. She lay so still, like the experimental aircraft she was, so to speak (still on the secret list): covered completely in a sheet of soft parachute silk, which stretched down to the floor on both sides. But her silhouette gave the illusion of completeness—a whole, undismembered body of a corpse, woman, doll or whatever. "You said she was still in bits," I said and Marchant tittered with pleasure. "They are not completely joined up as yet for action, but I want to give you the illusion of how she's going to be by showing her off bit by bit, so you don't see the joins. The power isn't in yet, but I get some traction off another unit which enables us to check the whole flexion patterns of our fine plastic musculature. I plug her into a g-circuit." He performed some obscure evolutions in the corner, switched on powerful theater lights above the body, and beckoned me over with a shy grin, lifting as he did so the corner of the silk to reveal the face. It was extraordinary to find myself gazing down on the dead face of Iolanthe—so truthful a copy of the reality that I started with surprise even though I had been expecting something like this. But what really took me away was the perfection of that fresh and dewy skin. "Feel it," said Marchant. I put my finger to her cheek; "She's warm." Marchant laughed; "Of course she is, she's breathing, look now." The lips parted softly and a tiny furrow of preoccupation appeared on the serene brow. In her dream some small perplexity had surfaced here. It was skin, though, it was human flesh. Here she was, simply lying anesthetized upon an operating table. "Iolanthe!" I whispered and the lips parted as if to answer me, but she said nothing.

Marchant watched my confused excitement with a happy air of complacence. "Whisper again and she will awake," he said, and in an incoherent, uncomprehending sort of way I said: "Darling, wake up, it's Felix." For a moment nothing, and then the whole face seemed to draw a waking breath. The lids fluttered and very slowly opened. "Damn," said Marchant. "Said has taken out the eyes again for restitching. I forgot, sorry."

Simply having the appearance and memories of the dead actress was not enough. The project scientists agreed that in order to become Iolanthe, the simulacrum had to live her life, and be treated as though she *were* Iolanthe:

So the great work moved slowly forward toward launching day; it was arranged that Iolanthe should imagine herself to be waking in hospital after an operation, recovering from the anesthetic. Once dressed she would be moved into a small villa which had been furnished for her with her own possessions—Julian had acquired them all, furs, and ball gowns, and shoes and wigs. In other words, to give her reaction index and memory a chance to function normally, we would provide best ideal test-conditions in ideal surroundings. All around her would be the familiar furniture of her "real" life—her books and folios of film photos, her cherished watercolors by famous artists (careful investment: all film stars buy Braque). . . . There would, then, on the purely superficial plane, be very little to distinguish between Iolanthe dead and Iolanthe living. Except of course . . . the dummy would be living the "real" life of the screen goddess.

Instructions to the staff reinforce the argument:

Above all, nothing must be said in the presence of the dummy to suggest to her that she *is* one, that she is not real. She must not be made to doubt her own reality—because that might lead to some sort of memory collapse; whatever doubts she may eventually have must come out of her own memory fund and its natural increment.

Easy to say, of course, but the thing was that she was so damn real that it was difficult not to think of her as a "person" . . . already! And she was not walking and talking as yet—the acid test of her mock humanity.

An earlier novel by Olaf Stapledon, *Sirius* (1944), concerned the creation of a superintelligent dog. The scientist in the story adheres to the principle that social interactions will determine whether the full potential of an enhanced brain will develop:

One other consideration inclined him to choose the dog; and the fact that he took this point into account at all in the early stage of his work shows that he was even then toying with the idea of producing something more than a missing-link mind. He regarded the dog's temperament as on the whole more capable of development to the human level. If cats excelled in independence, dogs excelled in social awareness; and Trelone argued that only the social animal could make full use of its intelligence. The independence of the cat was not, after all, the independence of the socially aware creature asserting its individuality; it was merely the blind individualism that resulted from social obtuseness.

Again, the social setting is seen as all-important in the development of the human qualities in the animal:

> This animal must have as far as possible the same kind of psychological environment as their own baby. He told [his wife] of an American animal-psychologist and his wife who had brought up a baby chimpanzee in precisely the same conditions as their little girl. It was fed, clothed, cared for, exactly as the child; and with very interesting results. This, Thomas said, was not quite what he wanted for little Sirius, because one could not treat a puppy precisely as a baby without violating its nature. Its bodily organization was too different from the baby's. But what he did want was that Sirius should be brought up to feel himself the social equivalent of little Plaxy [their daughter]. Differences of treatment must never suggest differences of biological or social rank.

Hiroaki Kitano, head of Kitano Symbiotic Systems, a private, nonprofit robotics firm in Tokyo, runs a robo-soccer team, and sounds a similar note. He believes a person's behavior toward a robot is very strongly influenced by its physical characteristics.

INTERVIEW WITH HIROAKI KITANO

Q: What is the most important element of designing a robot?

A: Designing a robot differs from making conventional customer electronics equipment. Industrial products are designed to be objects, whereas in robotics you will have to design the existence or the process that the robot will exhibit. Because the shape of the robot usually resembles animals or human beings, people project onto it a general conception of animal or human. Even if the shape is very different from existing animals, because it moves autonomously with a lot of moving components, resembling arms and legs and heads, people start projecting animal-like behavior and expectation into the robots.

So we have to be very careful in designing the shape and the behavior, so people will not expect too much from it.

When we designed our upper torso humanoid robot, we carefully chose a design that resembles an industrial product instead of coming too close to the humanlike appearance. We didn't use a silicon skin or any soft material, so that it doesn't really look like a crude copy of a human being. People should not think we are trying to copy animals or human beings.

Q : Why not?

A : The goal of robotics research is twofold: We are trying to create engineering artifacts that can be used for helping human beings, not trying to take over civilization. Secondly, through robotics people can understand the wonderful mechanics of human beings and animals. We don't want to trigger robot-phobia.

Q : What if somebody designs a robot to look like a human being?

A : The technology we have is no way near a stage that we are going to have any serious conflict between robotic systems and human beings. Robotic systems are very fragile at this moment.

Robots will soon do a whole lot of good things, very efficiently, given a specific situation, but there is no comparison with human beings. If someone comes up with the design and robotics which closely resemble a human and its behavior I think it will raise very interesting and intense discussions. And I think at some point that will happen and there will be issues that need to be raised then.

Q : In your own research you are working on the robot's interior, but people look at the exterior and wonder what you are doing.

A : The choice of design depends on our own definition of what the robotic system should be, what robotic systems can be within the context of our civilization and the relationship between men and the machines. At this moment robots are generally considered as tools, not their own species of whatever. That is why we chose to make our robots look like industrial products.

Others may chose to have something more closely resembling actual animals or human beings. I am not quite sure at this moment how people will react to that—whether they will get used to it, or simply reject it. My guess is, regardless if people like it or not, companies will start producing it. And if it appears harmless, people will get used to it, or accept that such things exist, anyway.

Robo-phobia in Media

What is the source of robo-phobia? Do people really believe superintelligent robots will challenge humans for control of the planet? That's Hollywood's frequent take on the future, at least in the high-budget, well-known films that are likely to influence the public discussion. (Low-budget films feature menacing robots, too, but more often menacing organic aliens.)

In *The Day the Earth Stood Still* (1951), the great robot Gort, which could annihilate Earth, is clearly under the control of its alien but very human-looking master, Klaatu.

In the comforting, vaguely medieval *Star Wars* empire, robots are mechanical helpers, owned, without rights. All conflicts occur between humans, or organic life-forms. Harking back to the 1953 *War of the Worlds,* huge, evil battle-droids that look like giant animals are merely mechanical vehicles driven by organic life-forms. Not even aliens are any smarter than people, and certainly not any machine.

In the 1980s TV series *The Hitchhiker's Guide to the Galaxy* and the film of 2005, Marvin, the paranoid android, with a "brain the size of a planet," is cheerfully sacrificed by a trio of humans without a second thought. Doomed to plunge into a star on an out-of-control spaceship, the robot grumbles, but doesn't try to countermand his organic masters.

All these robots are autonomous, self-aware, intelligent creatures, yet none is a threat to organic life.

The closest to an artificial intelligence gone amok is HAL, the self-aware computer running the spaceship *Discovery* in the 1968 film *2001.* During the trip to the outer solar system, it becomes deranged and has to be deactivated. In the next film, *2010,* it is explained that HAL discovered it had been lied to and felt this compromised the mission. HAL could not deal with the very human use of falsehood.

The *Terminator* films deployed robots that humans had

designed, but were coopted by the cybernetic system running the U.S. defenses. The seat of hostility to humans was then the newborn AI, not the robots. Significantly, the first *Terminator* robot (Arnold Schwarzenegger) returns in the sequels as a defender of humanity.

Polls find little support for a pervasive fear of robots by the public. They do not seem to have picked up the hostility that has many examples in science fiction. In fact, evil in stories often comes from humans or other organic life. Yet the threat of a cultural backlash seems very real to the robot designers, perhaps because they have read a lot of fiction of the post-Asimov generation, which did not see robots as beings born without "sin."

The following interview with Dr. Foerst on the cultural and historical antecedents of robots suggests that robo-phobia emerges from an earlier tradition.

INTERVIEW WITH DR. ANNE FOERST

Q: You deal with legends all the time. Can you give us some perspective on famous legends about artificial humans, like the golem or Frankenstein?

A: Humans have for many millennia dreamt of rebuilding themselves, and the idea was always to understand yourself. Every religion with an afterlife theory has something that survives death or is resurrected, that is interpreted as the essence of what it means to be human. It is the same thing with ideas about androids or homunculi. So they usually were disembodied, or they were just bodies.

We can trace the idea of humans rebuilding themselves as perfect machines back to Egypt and Greece between two thousand and three thousand years ago. So it is a very, very old legend. More recently, besides this century's sci fi, classics of rebuilding ourselves come from the Jewish tradition from Eastern Europe. These are the golem stories and the Frankenstein stories, built on Mary Shelley's novel *Frankenstein*.

Q: Is the golem old as well?

A: The golem stories go back to Psalm 139, which is the story of God being with humans all the time and having observed them before they were even born. In the Hebrew text, "Golem you did create me. You knew me already in the womb of my mother." This is the only time where "golem" appears, and there is no real explanation, except that there is a relationship between this kind of cluster of random self and God.

The original golem story talks about a real historical figure, Rabbi Judah Loew ben Bezalel, the Maharal, or highest rabbi, in the Prague ghetto in the fifteenth century. This was when the first pogroms started against the Jews by Christians. He had a theory that the world is completely made up of numbers and letters—which are in Hebrew the same—and that if you trace the order of the universe with those numbers and letters, you can rebuild everything you want. So he made a golem, a human body out of clay, and then chanted around him the order of the universe, which is God's reason. The golem then came to life. But the ultimate way for the golem to become alive was by putting the name of God either on his forehead or on a piece of paper on his tongue. Then it could participate in the spirit of God. The golem was created mainly to protect the Jews against the pogroms of the Christians.

As an example, the Christians would hide stillborn babies in the ghetto and then would come back the next morning with the police and say, "See, Jews kill babies in their rituals." So the golem would go out and take those stillborn babies away, so that there was no reason to accuse Jews.

Q: What is the upshot of this? Some people thought it was sacrilegious.

A: The interesting thing for me is the golem as prayer—God has created us in God's image, so we participate in God's creativity. Whenever we are creative, therefore, we celebrate God. But we are the highest beings of creation, so when we actually rebuild ourselves we are praising God the most.

I see that as a direct link to the lab here. We are trying to build robots—an analogy to human infants, and it is amazing how far we have to go. We become very modest as to our capabilities because even a newborn, which is fairly primitive, is so much better than any robot you can build right now. It is very hard to rebuild even parts of us.

Q: How did the golem story get its negative connotation?

A: There are two endings to the golem story. The rabbi would always take the slip of paper out of the mouth of the golem on the Sabbath so the golem would keep the Sabbath as well. But one Sabbath he forgot. The rabbi went to the synagogue and the golem was without a lord and a master—and went berserk. They get the rabbi out of the synagogue, he comes running and fights with the golem to get the slip of paper. One story is that he gets the slip of paper, and the immobile golem crushes him, and so they both die. The classic element of hubris: we can't control our own creations, and they might actually turn against us. This is a motif you can find all over science fiction.

In the other ending, which I think is much more interesting, the rabbi after a fight takes the slip of paper out of the golem's mouth, then takes the dead body and stashes it in the attic of the synagogue in Prague. And so, the legend goes, there it is till now. He told his sons, only his sons, the formula to revive the golem— numbers and formula. Many of the people who founded artificial intelligence (AI) come from that tradition, and have been told the formula to revive the golem. They are here right now in this building at MIT, so there is a direct link from that tradition to the founding and the beginnings of artificial intelligence.

Q: How does culture differ among the robot builders?

A: It's an intriguing idea that the fear of cultural backlash is stronger in the robot-building community than in the culture at large. The great science fiction writer Isaac Asimov, also from the cultural tradition of the golem, invented the Three Laws of Robotics to control the fictional positronic brains in his *I, Robot* stories. He did not assume that hostility was innate in intelligence.

Before we can dismiss this fear as just an abundance of caution, we should not forget that two other twentieth-century science fictional technologies with great promise ran into cultural backlashes: nuclear power and genetic engineering. Interestingly, the resistance was in different countries. Europe, as a whole, is much more accepting of nuclear power than the U.S., but it's the reverse for genetic engineering.

Q: We have not mentioned the Frankenstein story, although it is seen as a "creation gone wrong" story, because the monster is not a robot. What is your take on this story that relates back to robots?

A: The golem tradition is much more positive, has this kind of prayer element and spirituality element. The Frankenstein story is always perceived as negative because the monster ultimately turns against his creator and his whole family and kills them all. But I've never seen it that way. If we are thinking about what it means to be human—that is, to be part of a community, being embodied, using the body as a means to another end—then when we look at the treatment Frankenstein gave to his monster, it is clear why the monster had to turn against him.

He never got a name, he was never assigned personhood. Quite the opposite—Frankenstein (the monster) leaves and runs away immediately. He was shunned by all people. They never took him into the community because he looked horrid. The only guy who ever accepted him was blind. The monster was never able to become part of a human community. So where should the motivation of benevolence and niceness to other people come from? No wonder he had to turn against the community. I don't see that Frankenstein is an example of how our own creation turns against us—I see it more as an example of what happens to any creature when you treat him badly.

But there's another way. As long as robots remain small, they won't be seen as menacing. Industrial robots are often bulky, boxy, and have huge strength. Thousands of these assembly and welding machines labor twenty-four hours a day in auto plants, without incident; humans keep well clear of them, for safety.

Perhaps the safest way to avoid a cultural backlash against robots is to build them into nonthreatening bodies, like pets. The

Japanese have taken this approach from the first. After all, humans have a long history with pets, and we know how to relate to them. They're useful, or were when we needed hunting dogs and cats to police the mice, and we take care of them. That's not a bad model for a relationship with a household robot. It should imitate the responses of cats or dogs, move in animal ways, be much smaller than people, and when it must respond to people verbally, use ingratiating tones and phrases.

Already on the market, iRobot's home security robot, the LE, is about the size, and very roughly the shape, of a medium-size dog. Robo—vacuum cleaners, pool cleaners, and lawn mowers are the size and shape of large pond turtles. The hit toys of the 2000 Christmas season were robot dogs, with larger and more sophisticated ones to come.

None of them will get the newspaper for you from the front porch, but that may be just a matter of time. As they get smarter and more responsive, we will give them names, just as some people name their cars now. Someday, kicking your robot may be as socially unacceptable as kicking your dog.

iii

○   ○   ○   ○   ○   ○   ○   ○   ○   ○   ○

'Bots, 'Borgs, Bionics, and Betters

It is better to know some of the questions
than all of the answers.

—James Thurber

Reservations Recommended

The future promises a wealth of humanlike machines and machinelike humans. Along with them will come many misgivings about this latest expansion of the human prospect. We can expect the social battle to be joined over issues of rights, "naturalness," sexuality, and the very definition of human itself.

Today several trends seem to promise convergence in a future with no strict boundaries between the personal and the public, the natural and the artificial. In this part of the book we look at wedding computers to ourselves ever more intimately, both in clothing and implants, as well as through the global net that embraces us electronically, merging cultures as never before.

Does this adaptation portend a kind of being we have not envisioned? Our popular media already use the image of machinelike enforcers of the law, beings whose purpose, bodies, and will loom larger than those of mere fragile humans. Or is this portrayal a robo-copout?

In 1947, the writer Jack Williamson published "With Folded

Hands," a story that foresaw robots termed "humanoids, the perfect mechanicals" with the motto, "To Serve and Obey, and Guard Men from Harm." Gradually they maximize human happiness and safety by not letting us do anything remotely dangerous, like drive our own cars. At the chilling conclusion, a man who disliked what this was doing to humanity is lobotomized, erasing his objections and rendering him perfectly, mindlessly happy—and the humanoids think this lies perfectly within their guidelines. This theme was taken over bodily into the film *I, Robot*. The film ostensibly came from the famous Asimov stories, but its ending directly copies Williamson's novel.

Still, such a future seems unlikely; one doubts that any robots could suppress a true armed revolt by the most dangerous animal the world has ever seen—us. The entire *Terminator* series is based on this idea, but it seems quite improbable in view of humans' adaptability and creativity.

But humanlike physical similarity throws off human wariness, as the Kismet robot at MIT shows. We read facial features in machines and take them as social signatures. Perhaps a truly human-seeming robot will be able to exploit our natural affinity and social signals. Such ideas quite plausibly pervade a lot of science fiction.

Indeed, will the very distinction organizing this book itself, between cyborgs and robots, seem less important or even undefined a mere century from now? This part takes up the blending of these two modes, driven by the coming profusion of worn computers and the growth of the Internet. These will soon become symbiotic, leading to a synergy between men and machines that few anticipated.

CHAPTER 9

Wearable Computers: Blending Man and Machine

You are what you wear.

For children just being born, computers will not mean screens and bulky keyboards.

They will be fashion statements that do a job, just as shoes and hats do. Then they will alter our social lives, even our sense of what's personal.

Several companies—giant IBM, midsize Dallas Semiconductor and NCR, and tiny Charmed Technology among them—are developing lines of "smart" jewelry and watches that carry tiny speakers, microphones, Steve Mann's "personal digital assistants," or mouse-type peripherals. So-called digital jewelry could revolutionize social relationships. "You'd have to be very careful at a dinner party who you talk to," said Wallace Steiner of Tiffany and Company, in a recent newspaper article. It will all be there, retrievable.

This isn't just about expensive jewelry for the rich. After the first wave of wearables, other uses are already appearing in the commercial world. Historically, most innovations are adopted for business before the home, if they can save time or improve performance. It is the same now—wearables can deliver critical information on demand to busy people who need free hands.

Walkaround Wearables

Workers can now use compact computers that are rugged enough to go where they do—repairing factory equipment, inspecting finished cars, doing quick repairs, or keeping track of shipments. On assembly lines they fit in tight spots where a laptop computer is too big or fragile to work. A foreman can search a database and call up drawings of a new assembly routine. On a small display screen, perhaps on a head-mounted boom, a small color picture seems to hang before a worker's eye. Better, it can appear on the inside of interactive glasses—spex, some call them—overlaid on the left eye's field of view.

Physicians walking rounds can access patients' charts, get fresh diagnoses, and check e-mail updates without losing their momentum by having to duck into a workstation, assuming they're even near one.

Travelers popping open a laptop in a cramped airline seat might appreciate computers that can take orders from one-handed typing or, better, voice command—with quickly adaptable noise filters that can pick up their owners' voices even in the middle of jet engine whine.

Once, wearable computers were handcrafted and one of a kind. Now mass-marketing is about to make them available off the rack. Soon, they may be nearly unnoticeable. Expect them to catch on when they seem to fit naturally, leaving the user at ease. That's when they will blend into the world of fashion.

In recent models, the wearer has the illusion of reading a normal-size desktop screen an arm's length away. In fact the image reflects on the inner surface of eyeglasses. Commands are typed into a small keyboard worn on the wrist, or spoken into a small head-mounted microphone. A video camera picks up nearby visual information and sends it directly into the computing system, which rides in a fanny pack.

Wearables now range from five to ten thousand dollars, though experience with personal computers suggests that prices should drop within a few years. They may still be worth the high prices for employers, though, because they can add greatly to job efficiency.

Workers at Areva, of Lynchburg, Virginia, inspect steam generators in nuclear power plants wearing the five-thousand-dollar Xybernaut Mobile Assistant V underneath a radioactivity-shielded containment suit. By using a one-handed keyboard strapped to a wrist, inspectors maintain the inventory of test equipment as instruments come and go. Technicians pass a scanner over equipment as it passes.

The Bath Iron Works in Bath, Maine, a shipbuilder and refurbisher, has inspectors wearing cameras and a wireless phone link. These send digital pictures of trouble spots on the ship to a site near the yard. Engineers review the photos and recommend repairs, saving a lot of steps. Similarly, Northwest Airlines inspectors and mechanics document aircraft repairs on the spot, avoiding the traditional slow, in-triplicate paper trail synonymous with bureaucratic detail. Parts numbers and repair sites are entered quickly by bar code scanners.

Since the 1950s white-collar professionals in offices have been the most wired people. But because the big service corporations like shipping and telephone companies and car rental firms are adopting these technologies for their field personnel, a lot of blue-collar workers are going to be wearing cutting-edge wireless technology, too. Advanced wearables are being introduced to people who couldn't otherwise buy them, an unintended bottom-up marketing strategy.

Innovation from the Bottom Up

In the United States, new technology is typically scarce and expensive—"toys for the rich" at first, then later, when the price drops, available to all. This is the top-down marketing we are familiar with. But Japan follows a different paradigm: new gadgets are introduced by the millions at a low price that encourages everyone to buy them. Adoption of wireless technology in the United States may follow a quicker, more Japanese model. Cell phones have shown this over the past decade; in 2005 the United States ranked sixteenth globally in percentage of land covered by wireless phones.

Wearables are now at a level analogous to personal computers in the late 1970s, when small start-ups like Apple Computer and Microsoft were defining the standards for an industry. But wearables can draw on infrastructures from today's laptops, wireless handheld devices like BlackBerrys, and other portables that will speed wearable development. Batteries in wearables are the same lithium-ion cells as in notebooks, and the hard drives are also the same.

Fresh technologies, some quite unsettling, are on the horizon. Already users are tapping the power of their own bodies by using compression pads in their shoes, to feed electrical currents upward for storage in the batteries. Plans are afoot to use body heat to charge batteries. While power levels are low, the source is free and endless.

A wearable's hands-free operation drives voice-control to new heights. Present processing chips make users wait many seconds after some voice commands because of processing time, not merely the problems of deciphering speech. Even the most comfortable, trained software that recognizes its master's voice cannot yet understand very complex instructions, and sometimes need a hand-typed helper.

Despite the TV commercial of a young man sitting on a bench in Venice, saying "Gimme soybeans, scroll up, up, yeah, yeah, yeah, buy it, buy it, buyyyy it!" that wouldn't work with any actual wearable. Voice input–only wearables can filter out a steady background drone, but they don't work well in a noisy public setting, such as taking notes at a meeting. For those situations, there are keyboards strapped to the user's wrist available, pioneered for U.S. Special Forces troops.

To the wearer, most displays appear to hang in the air, and as with ordinary television, there is a trade-off between brightness and low power use, high resolution and large apparent size. Still, most wearables produce bright, full-color images easily visible in full sunlight.

Wearers stress that they would rather that people nearby not be able to tell they are using the computer at all. Wearing an unnoticed computer can also be important for certain users—the visually challenged, electronic media reporters working undercover, celebrities whose sunglasses shield a combined display and camera, or for anyone who wants to blend in.

Soon enough, these users may have laser diodes that splash images directly on the retina, so bright they can be made out even outdoors at high noon. That would eliminate the hanging-mirror giveaway. The coming compact wireless links, using microwaves or infrared sensors, could eliminate the inconvenient, techno-nerd dangling cables between the central processor and the microphone, keyboard, head-mounted display, and other peripherals.

The Wearable Frontiersmen

"Persistence" is Georgia Institute of Technology's assistant professor Thad Starner's term for being able and willing to wear a computer all the time without tiring. He ranks among the frontiersmen (and -women) of wearables. Starner wears his handmade system continuously while awake, the system constantly available to remember the names of people he meets, record conversations, or call up the Internet—all while he keeps eye contact with colleagues and students. "It sure beats taking notes."

The most persistent wearable pioneer is Steve Mann, assistant professor at the University of Toronto. He has worn a succession of his own wearable creations since he was a graduate student at MIT in the 1970s. He also stresses the benefits of looking "normal" while wearing a computer.

Within a decade, wearables will be part of one's clothing, fashion statements made practical, and integrated with ever-more compliant software. This will be the first step toward integrating people continuously and smoothly into a larger web of communications and information, with possibly far-reaching implications.

In 1998, Mann said:

Wearable computing facilitates a new form of human–computer interaction comprising a small body-worn computer (e.g., user-programmable device) that is always on and always ready and accessible. In this regard, the new computational framework differs from that of handheld devices, laptop computers, and personal digital assistants (PDAs). The "always ready" capability leads to a new form of synergy between human and computer, characterized by long-term adaptation through constancy of user-interface.

INTERVIEW WITH STEVE MANN

Q: Does a wearable computer mean a bulky machine around my waist?

A: The machine I've got now is fairly covert. The only giveaway is that cable, but if my hair is long enough, I can drop it down inside my shirt from the back.

Q: Covert?

A: "Normal-looking" would be the right word. The processor fits in a shirt pocket, the eye interactor doesn't look too strange. Over the last twenty years I have shrunk it down from the very large cumbersome devices of the 1970s. Now it's very small and light-weight.

Q: Is it important that it be designed to look normal?

A: Yes, it is. Face-to-face interactions with people are very difficult for someone wearing a large, obvious device.

Q: How is this technology going to evolve?

A: I am working on a smaller, sleeker, slender version now. One of the key new concepts is the vitrionic contact lens. Vitrionics is electronics in glass, and putting different kinds of things like optic elements and so on into small sizes.

   The eye interface absorbs and measures incoming rays of light, and then through communication with the processor converts them. Then new light is generated that goes into the eye. This is a "reality mediator system." The system augments, diminishes, or otherwise alters the visual perception of reality by intercepting and processing the rays of light as they pass through the device and before they reach the wearer's eye.

   The idea of *mediated reality* is to divert rays of light that would otherwise enter the eye, and to process, measure, and quantify these rays of light and then resynthesize rays of what I call virtual light. The virtual light comes out the other side, and goes into the eye.

Q : How does it send them into your eye?

A : A small laser light source is concealed in the eyeglasses and controlled by the computer. It resynthesizes rays of light.

The input side becomes much like a camera, measuring the light that comes in. When the camera is located on the eye, the eye in effect becomes the camera.

And then the other side of it is what I call an aeromac—a device that resynthesizes the rays of light coming out the other side. One of the key features is the zero-eye-strain reality mediator. The rays of light are not focused anywhere in particular, so it doesn't force the eye to try to focus at any particular distance.

Q : How does it put the image onto your eye?

A : Rays of light shine through the center of the lens of the eye wherein the light is not bent or changed, so that it has a sufficient depth of focus. One of the key concepts in the aeromac is the depth of focus control. Much like a camera has an aperture that allows it to have sufficient depth of focus to capture subject matter over a wide range, the aeromac also has a sufficient depth of focus to present the information to the eye over a wide focus range. It doesn't cause the eye to focus at a distant plane.

That means that somebody who normally wears prescription eyeglasses can take them off and see just as well through these glasses as with their prescription lenses on. In fact, I found that people with poor eyesight often see better through this type of high-tech device.

Q : It shines the reconstituted light right into your retina?

A : Well, no, it shines rays of light through the center of the lens of your eye. That's what I refer to as virtual light or converging rays of light.

A normal display will cause the eye to focus somewhere, whereas the purpose of this invention is to either allow the eye to focus wherever it would by its own course of action or to present the information on the same depth plane as subject matter in the scene.

Q : So the result is to put an overlay on the field of view?

A : You can either overlay or underlay or add or delete subject matter in the field of view. The "wearable face recognizer" inserts a virtual name tag, the "video orbit tracker" allows other subject matter to appear on top of the actual subject matter, or you can delete material. That's particularly useful with billboards and other visual detritus that invades our personal space. I refer to this as "real world spam" and it can be deleted from your visual field if you need to make room for other material. If you are driving to your friend's house you might see messages on these billboards that become directions on how to get there—customized messages, that only you see.

Or we can have customized messages that a small community sees—shared messages. Like leaving a message for my wife on the front of a shop, saying, "I was here, check out this special." So that message can replace spam.

We call this a packet filter. Packets of light come through, and the eyeglasses forward some of them to the eye and block others. We can think of it as a photonic "fire wall" that provides a level of security over the resource which is your brain's ability to process information. If somebody got into your home computer and caused it to execute instructions and tie up the CPU, you might refer to that as theft of computing resources. How is that different from spam that occupies our brain?

Take these signs at the side of the road—they are made to really attract attention. For example, I have seen a sign that says RED and is octagon-shaped, right at the side of the road like a STOP sign. When you are driving along you tend to notice it, but it is only an ad for some kind of telephone service.

This spam is cleverly crafted to get your attention, and what we really need, since the eye is the window to the soul, are shades on that window.

Clothing-Based Computers and the Networked Community

Q : What about computers in clothes?

A : The original clothing-based computer back in 1982 was built into the fabric as a kind of fashion accessory. Then I organized a number of sort of fashion shows. Later on I moved back to the original reason for all of this, which was to have a *personal* imaging system and be able to modify my visual perception of reality with the eyeglasses connected to the clothing.

The computer in the clothing is really for creating signals, for running the eyeglasses, collecting information from the eyewear and then sending it back into the eyewear again. That's where it is all heading.

These fun little fashion accessories you read about in slick media have no real practical value above and beyond what you would get if you just had a cell phone in your pocket. It is really a convenience item.

Back in the 1980s I built computers in clothing to help the blind, using radar systems, so the wearer could navigate and "feel" objects pushing against their body at a distance. We are still doing a lot of projects now with the blind vision system, more meaningful than something that is simply convenient.

You have to ask yourself is there a key inventive step that moves us towards being able to do something you couldn't otherwise do, or is it merely a convenience item. There's a big distinction between a gimmicky sort of convenience item that attracts a lot of press, and something that improves the quality of a person's life.

Q: What does it all amount to? Are these just fun gadgets, or something better?

A: There is a lot of hype lately, all this fashion fluff and just the general hype that I think may ruin the field in the same way that artificial intelligence and virtual reality failed. The common thing was the hype. Someone was telling me that Digital Equipment Corporation referred to their machine as a portable, a data processor, as opposed to a computer because the computer had been overhyped.

And I think we are starting to get to that point with the wearables—a lot of hogwash and very, very little substance. It's bad for the field because people need to refer to this as wearable technology or personal technology, and you would almost want to avoid the term "wearables" because it has that sense of lacking in substance.

Q: What is the substance?

A: The fundamental issue here is personal empowerment.

Q: How can that happen?

A: One area is assisting the visually challenged—this technology can actually make a difference to them.

Another area is in the notion of personal empowerment. We have all around us invasive technology like cell phones, which give us a certain freedom but they also take away a certain freedom. I have heard people liken cell phones to handcuffs. And then all these invasive technologies—"smart" floors, "smart" light switches, "smart" elevators, "smart" toilets, "smart" rooms—all this technology encroaching upon us. Cameras and microphones everywhere watching us, allegedly to make the world a better place, but in actual fact to make the world a better place for the architects of this surveillance superhighway. That seems to be a fundamental point that nobody else has clued into.

Q: What is this surveillance superhighway? Is your wearable technology any kind of answer to it?

A: Well in many ways this invention is taming the monster with a piece of itself. We have this technological infrastructure kind of caving in on us, getting closer and more into our everyday lives, with all of these various forms of infrastructure that we don't necessarily control.

Q: Such as?

A: Cell phones that track where we are and monitor our conversation. And things like clipper chips that monitor and censor your TV and Internet access. What we wind up with really moves us further from community in the true sense. People interact less with one another than with the representatives of the surveillance superhighway, the "guards" (as per Jeremy Bentham and Michel Foucault—social philosophers). And so it moves us more towards the prison characterized in Foucault's writings, a world in which there is a central guard facility watching over everyone.

Q: Would you describe your device as something that knocks the door down?

A: The goal is control of technology by the individual, to create a little bit of personal space. We can interact with others, decide to filter out material, ads, spam, create personalized messages and interactions.

Q: How does this technology enhance community?

A: I've been exploring a small community of users of personal cybernetic technology, and what happens when these people interact with one another. I think it is quite different and interesting because people make this world on their own. It is existential to create one's own world.

I'm talking about a small group of students and friends—a community—recreating and rebuilding their own world, amid this surveillance superhighway. I am not saying we should go back to the Stone Age when there were no surveillance cameras, but I am saying we should move forward in a new direction, a different future than the one that is being rammed down our throats by the people talking about "smart" rooms and intelligent environments.

Q: Could it go the other way, making you even more reachable?

A: It's certainly possible that the technology could represent an even smaller prison cell. No prison cell can be smaller than your own clothing. So if it is closed in upon us with centralized infrastructure and that kind of thing, it could be just as bad. Like any other invention, it can have good or bad uses.

Q: But what you are trying to do is tip the balance towards the individual.

A: Yes.

WearComps and Personal Freedom

Q: Let's go a big step further. Do you think computers will be implanted into humans?

A: That's a possibility I've been working on. One of the things I've emphasized to my students is what the individual can build easily in his own space.

In other words, I have tried to make instructions on how to build these systems. If you go to www.wearcam.org or www .eyetap.org Web sites, they describe this technology and include instructions so people can make these things themselves. So long as it is fashioned by the individual, it becomes much more in the individual's space.

Q : Would mass marketing of these devices send it in the other direction?

A : It could. It depends on what the market wants and if the market doesn't ask for the right thing, a lot of times it gets garbage.

Now it is interesting that a community exists on the Internet, the Linux movement [open source software, free to any user], which has been the antidote to a lot of this market garbage. And in a sense Linux is an example of creating one's own world, and having this coexist with the market forces. So there is a business model that works to allow the individual to control their own space.

Traditional business models try to dominate the individual and make them dependent upon the organization. It's the "drug model" of having you become addicted to the software so that you need to purchase upgrades from a specific single vendor.

And I think the alternative is the self-made Linux metaphor.

Q : What is different about eye-tap versus other wearables?

A : Well, the wristwatch video conferencing system I built is an example of a WearComp that is not an eye-tap device. It is a computer and it is worn on the body, but it is different from the eye-tap because it doesn't have a focus liberator and it doesn't augment reality. The eye-tap gets right inside your eye, so that you can augment or deliberately diminish the perception of reality. You can add new material but also filter out what you don't want to see.

Q : What about e-mail?

A : I used to respond to e-mail continuously. I could be walking across the street, get a message, and as soon as I got to the other side of the road, stand there on the sidewalk to reply to it. I developed a reputation for replying within fifteen or twenty seconds.

So people e-mailed me a lot. But that didn't work out so well because it ate into my life excessively. To me it's just a side effect, but people often ask if I can read e-mail and surf the Web with my wearable, and the answer is that I've been clicking text for more than twenty years now with this kind of machine.

Q : What percentage of your communication is electronic versus voice?

A : Well this is also voice and video—the interesting thing about this invention is it blurs the distinction between cyberspace and the real world and between electronic communications and mediated communications. As I am looking through the machine seeing the world, somebody else can share that world. My right eye is tapped, so it becomes a shared space on the Internet, in both directions. As you would tap a telephone to get both sides of the conversation, when you tap my eye you can read out and write into. So people can see what I am looking at and they can also write into my eyespace if they have writing privilege on my eye. This then forms a communication space. It's not traditional video conferencing, which has failed in the marketplace because it doesn't add much to see a picture of me.

What my wife and friends and relatives want to do is get inside my head and see the world from my point of view. With eye-tap you can "beam me" rather than just "see me."

Q : That brings to mind the 1999 movie *Being John Malkovich*: the characters find themselves inside his head, looking out through his eyes.

A : Interesting idea. I didn't see it, but if I am at the grocery store, my wife at home can look out through my eye, draw an X through one thing and circle something else with an arrow over it, to direct that viewpoint. Usually I am also talking to her while she is animating the visual space, and this forms a much more powerful interaction than video conferencing.

Q : Is this a form of collective consciousness?

A : Well, from time to time I've recognized people I've never met before, because somebody on my Web site looking out through my eye sends me a message saying, "Please say hello to the person standing in front of you, who is an old high school buddy of mine." So this touches on the whole notion of a collective consciousness.

Q : Do you think you have become dependent on it?

A : People who have never worn shoes have very tough feet, but once you wear shoes, you lose some of your ability to adapt in the wilderness. Many of us if naked in the wilderness might not last as long as people who were normally naked and adapted to it. So we could say that clothing is harmful because it has built up a dependence.

Now we have pockets to put pens and notepads in, we can read and write, and we have become computer literate. As a result of all these technologies, we have evolved into a different lifestyle. The question is often asked, should we go back to that earlier lifestyle or should we be able to go back to it?

Calculators and computers have also made us soft. How many people now can calculate the square root of two, to a hundred decimal places, just with pencil and paper? It wouldn't be as automatic as in previous generations, but is that really a problem?

Are we simply evolving into this other life-form, and should we still learn how to survive in the wilderness, how to compute the square root of two, how to calculate pi, and if we were ever on a desert island should we all learn Morse code so that we could survive?

These are important questions and there may be some element of truth to these fears.

Q : Do you think there is going to be a market for the wearable technology, and it will be adopted?

A : I think some people will have it.

Q : How much of your world of communication with other people is mediated through the eye-tap?

A : Often I will be interacting with one person through the eye-tap together with other people interacting, so I have a local audience which is interacting directly through the machine and a remote audience which is remotely interacting through it. But it is really hard to quantify because the device has at least one mode which I call the "identity mode," where no input is getting changed. A lot of times the device just sits in that mode, always ready, running in the background.

Wearables face a major problem: many find them uncomfortable. It's no accident that the vast majority of wearers are guys. Of course, men are more likely to be techno-geeks. Beyond that, though, the wearables' wires and straps don't work well with women's clothing. And they're *heavy*.

Steven Feiner is professor of computer science at Columbia University with a research group at the Computer Graphics and User Interfaces Laboratory. He and his students work on augmented reality wearable computer systems, wearing the prototype backpacks and headgear they've created. Although the backpacks currently weigh in at about fifteen to twenty pounds, Feiner expects continued advances in miniaturization to shrink the computer to the size of a portable CD player that can be worn clipped to a belt. Like Steve Mann's and Thad Starner's groups, they are creating a networked community.

How far are we from off-the-shelf wearables for all of us at discounted prices? It's not just around the corner, but do-it-yourself kits for the technologically gifted are being offered by Charmed Technology of Beverly Hills. These consist of a belt-mounted CPU, one-handed keyboard, and a viewscreen mounted on eyeglasses. Charmed.com also offers a wrist-mounted device. Alex Lightman, CEO of the company and a graduate of MIT's Media Lab, organizes fashion shows of the wearables around the world.

Besides the now-standard head-mounted display (viewscreen), smart wrist computers are being developed by several different groups. Dick Tracy's "two-way wrist TV" may soon be on sale at your favorite computer store, offering you the ability to make notes, receive and send e-mail or voice mail, listen to music from an Internet site, be reminded of your appointments, and even find out what time it is.

The Wearable World

Assume these devices do become inexpensive and comfortable. What will it be like to live in a networked, wireless world?

Suppose you're a technophile. The main difference between today and say, about five years from now, is that all your devices will talk to each other (and they might run in Linux instead of Windows). If the wearer is interested in finding a particular movie, the video-screening program will sort through the week's TV offerings and pinpoint time and date, then relay this information to the wearer and the recording device.

In the future, the wearer will be cocooned in a personalized information space. Like the proverbial rose-colored glasses, the spex will filter the world for the benefit of the wearer, eliminating unwanted information and presenting a version of reality altered to suit the tastes of the wearer.

But the system will be able to do more, such as seeking out and retrieving wanted information from anywhere in the worldwide information network. If knowledge is power, ordinary people will become all-powerful.

In Wil McCarthy's 1998 novel, *Bloom,* the author describes life and relationships in a networked community. The protagonist and others wear "zee-specs"—wearable computers integrated into eyeglasses. They receive e-mails (VR mail in the book), their computers can record conversations and events, and the zee-specs are able to relay information to others.

> Even his zee-spec was an older model, blocky, folding his ears back, weighing on the bridge of his nose, leaving his features to sag that much more.
>
> "John Strasheim, hi," he said, rising from his chair and extending a hand. "Thanks for coming on such short notice. You're a few minutes early, actually."
>
> Shaking the proffered appendage, I shrugged. "Just eager to oblige, I guess. What can I—"
>
> "Take a seat, then. Set to receive a flash?"
>
> "Sure." Who wasn't?

His thick fingers danced in the space between us. My RECEIVING light went on, and the air before me came alive with information, image windows and text windows and schematic windows rastering in and then shrinking to icons as my spec compressed them in working memory. Too quick to see much in the way of detail. Pictures of blooms, I thought . . .

His fingers stroked the air, manipulating symbols and menus I couldn't see. One of my image icons began to flicker. I touched and expanded it, moved the resulting window to the lower right corner of my vision. It was a video loop, false-color, depicting a complex mycorum which replicated itself over and over again. . . .

"This other man is Tosca Lehne," Lottick said to me, flapping a hand in the air between us as if trying to form some invisible connection. Maybe doing something meaningful on his zee-spec, or maybe not. It's hard not to notice that these devices, so rarely removed, have brought us a whole new body language, have encouraged us more than ever to speak with our hands, to sketch invisible lines in the air whether or not they'll be turned into real lines in the specs of our audience.

Not every side effect of these technologies is appealing. The society in the novel has developed new rules of etiquette. These deal with novel social situations that come from a networked community, much as we are currently grappling with cell phone etiquette.

It's bad enough when the guy next to you in a bus or restaurant is talking to somebody by cell phone, in that loud, abrasive way some have. How about when he's got the whole world on the line, through the Internet?

And worse, how about that camera over there by the door? It's wired into the planet, too. Is that lens turning toward you right now, looking you over?

I See You: Lusting for Privacy in a Networked Community

Quite soon, in a networked world, we'll have to adjust to different concepts of privacy. In turn, this will collide with security.

One of the advantages of having people share the view from your eye-tap is that others can ensure your safety. This illustrates a general rule: a smart, wired world is less risky. And less private.

Before entering a dark street, a woman notifies her husband or a friend that she would like real-time monitoring. Scanning the lot, her spex record faces and license plates, and this information is shared with her monitors. In a parking lot, her system can interact with the surveillance digital cameras already in place in the lot with the same request. Such surveillance systems, originally deployed to safeguard teller's windows in banks, soon spread to ATMs, elevators, hotel lobbies, and hallways. Once limited to private property, they are increasingly being used outside, in public spaces like street intersections and parking lots, to spot traffic offenders and vandals.

Where does safety leave off and intrusive monitoring begin? That's going to be more of an issue in the future. Already, phone company technicians using the new wearable computers are wary of having global positioning system (GPS) capability added to their devices, as it would enable managers to track their movements in the field.

In Wil McCarthy's *Bloom,* the pace is frenetic. His people don't live in Manhattan, but it feels as though they do.

On my zee-spec, images threatened to crowd out the real world altogether: paired data gene sequences scrolling upward in tandem, the duplicate portions flashing like alarm lights; the Io Senegen and Innensburg mycora, pulsing with false-color image enhancements and shifting annotations from the library tutorial; a map of the solar system, with *Louis Pasteur*'s course charted out as a dotted white line swinging close by Mars, kissing the orbit of earth, and then finally rising back toward the Immunity, toward the cold and dark of the upper solar system.

And of course I had a media window cycling slowly and methodically through my own net channels as well. . . .

Beside me, Tosca Lehne snorted and banged a cup on the table. "Hey, Strasheim, she's talking to you."

"What?" I looked up, saw that Jenny Davenroy had been speaking to me from across the table. "I'm sorry, I didn't catch that."

Shaking loose a few strands of unruly, tin-colored hair, Davenroy rolled her eyes and stabbed pale fingers at the air. "I said, what are you *reading*? Pardon my nosiness, but what we read at the dinner table says an awful lot about us as people."

"Yeah," Tug Jinacio chipped in, "especially what we read when people are trying to meet us. Come on, give with it. Geek it over."

"It's nothing," I assured them. "Just a little homework."

"Flash it to me," Jinacio insisted, not quite rudely . . .

"What better way to get to know you than by rifling through your private thoughts? Do please allow us."

"Really, it's nothing," I insisted. . . . But I flashed them copies of my windows.

In the future, the hostess at a dinner party will raise a toast and remind people to turn off their earrings.

Our Private Past

It may come as a surprise that privacy is not universally defined, and what constitutes personal space versus public knowledge varies widely among cultures. The people debating privacy issues and the emerging networked community are, almost without exception, arguing for the kind of personal anonymity that is very recent in human societies, and arose only with the advent of cities.

In Western culture, parents value personal privacy, shielding children from the physical facts of adult relationships. Rafts of books have been written advising parents to move the children into their own rooms as soon as possible for the good of all parties.

But traditional Eskimo nomads lived in one-room summer tents or winter igloos, the whole family snuggled together under common furs. When they settled into permanent housing, anguished fathers smashed holes in the walls between their bedroom and those of the children.

In medieval times, most people in Europe lived in small villages, clusters of thatched-roof buildings in the middle of fields, or large households centering around great houses or castles. In these small communities, privacy as we know it was impossible. The fun of Shakespeare's *As You Like It* is the absurdity of pretending that people didn't know who was sleeping with whom when all were living together under one roof.

Even today, the many people worldwide living in small towns and agricultural communities share their lives with others in a multifaceted relationship the Germans named *Gemeineschaft*. This means that when you interact with someone in, say, the hardware store, you also know where she lives, what her hobbies are, and what her family is like. In the city *Gesellschaft* relationship, you know nothing about the hardware clerk except that she is filling that job.

In the big urban agglomerations, personal isolation is the biggest problem. There are masses of people, but they're all strangers. So clubs proliferate, providing smaller, interest-villages amidst the endless urban sprawl. With the advent of the Internet, it became possible to create virtual groups.

This is an example of a general rule: *People shape technologies to their own desires*. The architects of these systems envisioned a kind of hands-on telex system, with utility mainly for business transactions.

Such lack of foresight is common. The designers seldom know much about how people actually get by. What we are seeing instead is a reflection of the human need to re-form the small groups we feel most comfortable in. Most people are online to stay connected with friends and family, or to find new friends.

The networked community is the next step, enabling continuous communication, canceling the impact of geographical distance. In Europe and Asia, cell phone usage is ubiquitous, much more so than in the United States, although we are catching up. People are always connected.

Despite the way we live today, the anonymity of the big city is historically recent. The anomie treated by writers as varied as Kafka and Raymond Chandler may turn out to be a temporary phase in cultural development. Staying connected may outweigh privacy, having deep roots in our evolutionary past as a group animal. The fear of being cast out of the group has always been the harshest punishment: "shunning" in Amish society, excommunication by the Catholic Church, or solitary confinement in the penal system.

George Orwell's *1984* foresaw heavy-handed surveillance by the State, Big Brother, with no personal freedom, and that image colors much of the gathering public debate about privacy. But what people fear more, underlying the basic reason we tolerate intrusive media, is secrecy. Almost universally, humans prefer all to be out in the open. In a networked society, hiding information will be virtually impossible.

In 1996, Rob Hall, a New Zealand mountain climber, froze to death high up Mount Everest, but not before speaking to his pregnant wife at home in Australia. Even though he could not be rescued, he was not out of communication. Unlike legions of explorers before him, he did not die alone. His singular experience underscores the power of the modern electronic communications network to connect people.

The networked community may bring about a truly global village—not an endless shopping network, but connections between people on an unprecedented scale.

Technobuzz

The wearables on the market today, and those under development, are roughly similar, feeding information into the eyes and ears of the wearer. But think beyond that: Could wearables mean something other than wires in clothes? What about keying information through touch? Smell? Taste?

In Gregory Benford's 1996 short story, "Zoomers," the characters work in cyberspace in one version of an advanced, networked, information-dense world:

> The pod wrapped itself around her as tabs and inserts slid into place. This was the latest gear, a top of the line simulation suit immersed in a data-pod of beguiling comfort.
>
> Snug. Not a way to lounge, but to *fly*.
>
> She closed her eyes and let the sim-suit do its stuff.
>
> **May 16, 2046.** She liked to start in real-space. Less jarring.
>
> Images played directly upon her retina. The entrance protocol lifted her out of her Huntington Beach apartment and in a second she was zooming over rooftops, skating down the beach. Combers broke in soft white bands and red-suited surfers caught them in passing marriage.
>
> All piped down from a satellite view, of course, sharp and clear . . .

Get to work, Myung, her Foe called. *Sightsee later.*
"I'm running a deep search," she lied.
Sure.
"I'll spot you a hundred creds on the action," she shot back.
You're on. Big new market opening today. A hint of mockery?
"Where?" Today she was going to nail him, by God.
Right under our noses, the way I sniff it.
"In the county?"
Now, that would be telling.
Which meant he didn't know.
So: a hunt. Better than a day of shaving margins, at least.

Is this plausible? How far can we integrate the body with its sources of information?

The prospect of using all five senses to the maximum means that a torrent of data could in principle come flooding into us—in place of our normal perceptions, of course, at least momentarily.

Currently our paperwork-clogged offices use skills in unnaturally narrow and unforgiving channels. We evolved to live in complex visual, tactile, and social settings, alert to subtle opportunities or threats. Office workers work with a handful of simple symbols on a featureless field, often on a self-lit screen. And while a dropped berry is of little consequence to a gatherer, a dropped digit can invalidate a whole calculation.

So a future "office" worker will not feel as if she's sitting at a desk, information feeding through her fingers and into keys, finally washing up on a screen. No more worshipping at the keyboard!

Here's how the above action in "Zoomers" works out:

She and her Foe were zoomers, ferrets who made markets more efficient. Evolved far beyond the primitivo commodity traders of the late TwenCen, they moved fast, high-flying for competitive edge.

They zoomed through spaces wholly insubstantial, but that was irrelevant. Economic pattern-spaces were as tricky as mountain crevasses. And even hard cash just stood for an idea . . .

She sensed the county's incessant pulse, the throb of the Pacific Basin's hub, pivot point of the largest zonal economy on the planet.

Felt, not *saw.* Her chest was a map. Laguna Beach over her right nipple, Irvine over the left. Using neural plasticity, the primary sensory areas of her cortex "read" the county's electronic Mesh through her skin.

But this was not like antique, serial reading at all. No flat data here. No screens.

She relaxed. The trick was to *merge,* not just observe.

Far better for a chimpanzee-like species to take in the world through its evolved, body-wrapping neural bed.

Was that uneasy sensation natural to her, or a hint from her subsystems about a possible lowering of the prime rate?

Gotcha! the Foe sent.

Myung glanced at her running index. She was eleven hundred creds down!

So fast? How could—?

Then she felt it: dancing data-spikes in alarm-red, prickly on her left leg. The Foe had captured an early indicator. Which?

The hunt is lots more interesting than yet another office memo. . . .

Our peripheral vision and hearing which kept our ancestors alive is a mere distraction to a clerk. Yet these senses are essential to who we evolved to be.

Think of ourselves as data processors. Our eyes link so profoundly to our eyes that they are inseparable, taking a large fraction of our brains to manage. Megabytes per second of raw data streams through our eyes, ears, noses, and skins, overwhelming more than a computer can presently process. We filter and analyze this flood instantly. Each of the retina's million effective pixels, working closely with the optical cortex at the other end of the nerve feed, can pick out in an instant whether a boundary is moving or not, and the rest of the brain ponders whether to worry about the fact.

The excerpt above holds out the promise that an office worker need not focus on narrow lanes of information. Instead, the process could be as liberating, as physically stimulating, as a flight over a sunny, salty beach.

This freedom is crucial. Wedding our bodies to our machines can liberate us, not enslave us to a narrow spectrum of life. Cyborgs, robots, wearables, enhanced senses—these are parallel channels running down to a great opening ocean of possibilities.

Which of these will prove most critical? *All of them*—because they will endlessly merge, reform, mutate anew. This dance will never be over.

And the changes will be tailored by the people who have the most at stake. It is truly an old human theme, worked out anew: *freedom through tools.*

And beyond that?

10

Metaman

Superintelligence?

I believe that the magic solution to AI, insofar as there is one, is not robotics but—the Net.
Web robots and more sophisticated Internet agents, not physical robots, are the ideal bodies
for the first generation of real AI systems.

—Ben Goertzel

If we become a worldwide society with members having a range of skills, both artificial and natural, how will they blend? This calls forth from UCLA's visionary Gregory Stock a biological analogy:

The extraordinary moment we live in today can only be seen if you take an evolutionary perspective and look at the larger trajectory of life on earth. Life first appeared on the planet some 3.5 billion years ago as bacteria, which are essentially little sacks of biochemicals. Bacteria came together in a cooperative cluster to form something called a eukaryotic cell, which has a nucleus and all sorts of organelles, the small organized bodies with specific jobs to do. It's a very complex structure, maybe a million times the volume of a bacterium. Cells then came together to form multicellular organisms, which is all the animal and plant life around us. And now multicellular organisms, through humans as the vehicle, are binding together to form a superorganism that is planetary in scope, and that has all of the powers at another level from our own. Many of the things that we see as our most human aspects—language, culture, values—are really properties of this collective, larger entity that is now being able to achieve essentially a global mind.

This is Metaman, the Web-wedded entity shaping now under fresh forces: globalization, the Internet, and the digital revolution. Where can these forces lead, given that we ourselves will become increasingly a cyborged, android, robotic culture?

Stock is a dapper, intelligent-looking man, appearing quite practical. He directs the Program on Medicine, Technology, and Society at UCLA's School of Public Health. In this role he explores critical technologies poised to have large impacts on humanity's future, particularly on the shape of medical science. His ideas rove to the outer precincts of current thinking. He remarks,

> Right now we are witnessing an event of an evolutionary significance equivalent to single-celled organisms joining together to form multi-celled ones. We are aggregating into a superorganism glued together by technology, seen most organically in the Internet, which almost has a life of its own.
>
> This is just the beginning. Imagine where this is going to be a hundred years from now, or a thousand. That is just an instant in terms of evolutionary time, yet it will be enough to radically transform everything about life. So it is no surprise that the powers of this superorganism are now feeding back and reshaping who we are— changing our very selves.

One grasping of the possibilities for how we will interact with such a global entity is to continue on in the short story, "Zoomers" as we follow a woman linked to many perspectives, all coming to her through her full sensorium of eyes, sight, touch, taste, and hearing:

> Every day more water flowed in the air over southern California than streamed down the Mississippi. Rainfall projections changed driving conditions, affected tournament golf scores, altered yields of solar power, fed into agri-prod.

Down her back slid prickly-fresh commodity info, an itch she should scratch. A hint from her sniffer-programs? She willed a virtual finger to rub the tingling.

—and snapped back to real-space.

An ivory mist over Long Beach. Real, purpling water thunder-clouds scooting into San Juan Cap from the south.

Ah—virtual sports. The older the population got, the more leery of weather. They still wanted the zing of adventure, though. Through virtual feedback, creaky bodies could air-surf from twenty kilometers above the Grand Canyon. Or race alongside the few protected Great White sharks in the Catalina Preserve.

High-resolution Virtuality stimulated soft, lacy filigrees of electro-chem impulses throughout the cerebral cortex. Did it matter whether the induction came from the real thing or from the slippery arts of electronics?

Time for a bit of business.

Her prognosticator programs told her that with 0.87 probability, such oldies would cocoon-up across six states. So indoor virtual sports use, with electro-stim to zing the aging muscles, would rise in the next day.

She swiftly exercised options on five virtual sites, pouring in some of her reserve computational capacity. But the Foe had already harvested the plums there. Not much margin left.

She killed her simulated velocity and saw the layers of deals the Foe was making, counting on the coming storm to shift the odds by fractions. Enough contracts-of-the-moment processed, and profits added up. But you had to call the slant just right.

Trouble-sniffing subroutines pressed their electronic doubts upon her: a warning chill breeze across her brow. She waved it away.

She dove into the clouds of event-space. Her skin did the deals for her, working with software that verged on mammal-level intelli-gence itself. She wore her suites of artificial-intelligence . . . and in a real sense, they wore her.

She felt her creds—not credits so much as *credibilities,* the oper-ant currency in data-space—washing like hot air currents over her body.

Losses were chilling. She got cold feet, quite literally, when the San Onofre nuke piped up with a gush of clean power. A new substa-tion, coming on much earlier than SoCalEd had estimated.

That endangered her energy portfolio. A quick flick got her out of the electrical futures market altogether, before the world-wide Mesh caught on to the implications.

Up, away. Let the Foe pick up the last few percentage points. She flapped across the digital sky, capital taking wing.

She lofted to a ten-mile-high perspective. Global warming had already made the county's south-facing slopes into cactus and tough grasslands. Coastal sage still clung to the north-facing slopes, seeking cooler climes. All the coast was becoming a "fog desert" sustained by vapor from lukewarm ocean currents. Dikes held back the rising warm ocean from Newport to Long Beach.

Weather was now the hidden, wild-card lubricant of the world's economy. Tornado warnings were sent to street addresses, damage predictions shaded by the city block. So each neighborhood got its own rain forecast. The world was linked and smart and decidedly nonlinear, yes.

INTERVIEW WITH GREGORY STOCK

Q: Will this superorganism have a separate life of its own?

A: When you look at the aggregate of human activity and combine with that all of our creations, all of the technology that is in fact integrating it, this is an immense superorganism. It is essentially a creature of its own, and it has behaviors that are *emergent*, meaning they are not the expression of any individual or group of individuals. That is terrifying for many people but to me it is very reassuring. Because our biggest problems arose when individuals think they are smart enough to determine what the future of human society is going to be like. They designed very rigid rules, lacking the informational content available when you integrate the activities of countless human beings through the marketplace, through the Internet, all dynamically adjusting.

Q: Does this larger entity have an independent mind?

A: You can see that it has behaviors that are almost mental. When you walk down the street, you look off and you see a pretty woman or a tree, and your mind flits from thing to thing. Well, look at what is happening to [the global consciousness]. Suddenly we concentrate on the Arab–Israeli conflict, then we jump off on what is happening in the ex–Soviet Union, and then there's a shooting in a school—so there is this jumping attention. Everybody gets all worked up and then they forget.

That is very much a higher mental phenomenon, a system with self awareness. What else could you label it? Our understanding, for instance, of global warming—no individual could possibly be aware of half a degree of change of average global temperature over a century. This information derives from hundreds of thousands of individuals, machines, computers, and sensing devices, integrated and then offered to us. We take it as our own.

And it becomes something that we almost think of as an expression of our own intelligence, of our own wisdom. But it's really the shadow of this metaorganism, global in extent and becoming increasingly powerful.

Q: This is a common image in science fiction, and so is the usual plot development: Can you imagine it developing an agenda?

A: If you look down from space at the night lights in the world over time, as if it were time-lapse photography, you would see them extend and pull back and grow more complex, and it would look like an organic sort of an entity.

Certainly this entity will be driven by its own purposes, which are likely to expand out toward the stars. What else could it do to increase its cognitive powers? But that doesn't mean that its agenda would be in conflict with what we do as human beings, any more than an individual would feel his motivations could somehow be in conflict with those of a heart cell.

Q: In an Arthur C. Clarke story of the 1960s, the world telephone system reaches a complexity threshold and wakes up, becoming sentient. Greg Bear envisioned an intimately connected world in his novel, *Slant.* Will your Metaman Web be at all like us?

A: The cells in our body are supplied with all they need to function effectively; that's the signature of a well-adjusted organism. I think the same sort of thing is occurring in the larger superorganism. It is an expression of all humanity because it is really built in our image. Interestingly, sometimes we are surprised by what we want. People are shocked that there is an enormous amount of pornography on the Internet. It's there because we want it.

Q: Will the global organism that you speak of have some kind of other intelligence?

A: There is a sense that to have intelligence you need to have some form of embodiment. I don't think that really means you have to have a physical manifestation as much as that you need sensory inputs. The interaction between intelligence and sensation allows learning to occur. Sensory input allows control to occur, and I think that is absolutely necessary. But there will be less and less physicality in the intelligences that are created and evolve. This larger global structure is going to have all sorts of realms of extended intelligence that are then used in various ways.

Q: How will this show up locally?

A: For instance, a future robo-vac that cleans your floors would be able to download a lot of information about its location, or the experiences of other devices that do similar sorts of things, and so it will be an extended intelligence that can move in ways that we can't imagine very clearly because our own intelligence is physically embodied, and not downloadable. A lot of machine learning occurs in a physical form, but it can be downloaded and transferred to another entity. That is one reason the machine evolution can be so rapid—because all experience can then be copied and replicated. You can't do that rapidly in biology.

Q: Can this global superorganism develop an agenda separate from or beyond a democratic society?

A: What would be the motivations or the goals of a global entity that is essentially the aggregate of all human and computer activity, the Internet, all sorts of things? Well, clearly it transcends our lives. In the same way a single blood cell merrily moving along in the bloodstream is not going to be able to comprehend motivations like love and hunger: these sorts of things would have no meaning at that lower level. So the motivation of a superorganism to move out towards the stars and diffuse out into the universe can only happen on that large scale. This is not something that is going to happen on an individual basis.

Q: A different analogy would use a neuron, the elemental connection in the brain and nervous system. A single neuron cannot know the idea of love, but it can convey it along a chain immensely larger than itself.

A: Yes—this is something so far beyond our sphere of activity, we are doing well just to conceive of that entity. It requires all of our technology to allow us to do that. We only really started seriously thinking about these things when we looked down from space and could see the totality of human activity. When we are lost in narrow interactions, we really do not see this very well, we don't even think in those terms.

Q: This again recalls Arthur Clarke's vision of the satellite and phone system suddenly becoming conscious, an idea from forty years ago.

A: Suppose you talk about the time scale of activity for the larger superorganism, things that we are concerned about—maybe global warming or flooding—and you think in terms of centuries. Well, this larger entity just moves around, it withdraws, it populates an area. If we knew that Florida, for instance, were going to be under water in a hundred years, not only would this probably *not* have an enormous impact on our lives—because people would tend to just move—but the entity as a whole (the summation of human activity) would be unaffected by it. It would just go off to some other place and do much the same things.

But suppose it doesn't leave us alone? How to avoid pests?

The Downside of Smart Cities

Technologies never evolve in a vacuum. They must integrate with other advancing technologies, older ones, and most especially with the quirky people who use them. To focus on concrete, lived experience, let us try a short imaginative fiction:

A SCENARIO

He was walking into the mall when the side of the Macy's building said to him, "Hello, Albert! So happy to see you again."

A big, glossy white smile exploded across the crimson Macy's display wall. The pixels were old-fashioned, big blotchy ovals and squares, so the teeth wobbled and the lips jerked back and forth from red to scarlet. Practically antique. He couldn't recall having been here, a second-rate mall looking a bit run down, but the gushy wall rushed ahead in its cordial, silky woman's voice.

"You last graced us with your presence 2.43 years ago, when you purchased some camping equipment at Let's Go! one of our most popular stores."

"Oh yeah." He slowed. "I was just gonna pick up some shoes—"

"We're all very sorry, but Let's Go! has . . . gone. Left us."

"Out of business?"

"Sadly, yes. Customer demand for outdoor equipment has fallen. But!—" The voice brightened. "—we have a new store, ComfyFit. They have a wide, fashionable selection for big, athletic men like you."

He wasn't big or athletic, but he wasn't dumb, either. "Can the compliments."

"Oh, and assertive, too!" the womanly voice boomed happily. People passing nearby looked at the wall and snickered.

He hit his shutout control, but the mallwall went on babbling in full-color big-screen enthusiasm about the bargains "just a few steps away!" Multicolored maps flowed across the pixels, in case he was brain dead and needed guidance.

"Damn!" He walked faster. His inboard software was only three months old, but already even this cheapo mall could block his disabling defenses. Like most people these days, he was in a perpetual privacy battle with the invasive world. Lately he was losing. He made his way past the wall but the images followed him, splashing in gaudy crimsons and blues along the foyer of the mall itself.

The map highlighting the ComfyFit store led him to it, and he ducked inside before it could embarrass him further. But when he came out with a pair of what the Twen-Cen had called sneakers (did anybody sneak anymore?) a satiny voice said, "I'm so sorry about that, Albert."

"You should be." He didn't slow or even glance around. He could tell from the well-shaped tones that the mall had him on a tunnel mike rig, trapping their talk in a bubble a few feet wide.

"It's just that, you were my very first customer."

"Huh?"

"Think back 2.43 years ago. I was on my first outing, just a simple greeter program, building up experience. I hailed you and advised you about Let's Go! Don't you remember?"

"Sure, just like I remember all the traffic lights I go through."

"Oh, I like the way you say that. Almost like Bogart."

"Go away."

"This is going badly, isn't it? Believe me, I'd do anything to make it up to you."

Was that a Marilyn Monroe sigh? Sonic focusing was so good now, they could feed you anybody's voice. Probably the cameras embedded around here had profiled him, white-straight-unaccompanied-midrange affluent. "Okay, how about a discount on these shoes?"

"That transaction is complete," the voice said stiffly, like a schoolmarm, and then immediately, "Oh—sorry, that was the override program. I'll stop it—there!" The Marilyn voice came back. "Now I can arrange the discount, immediately."

His pace slowed. "Huh? You're two programs?"

"Discount done!" she cried happily. Then her tone shifted to close, husky, intimate. "Think of me as a person. A woman. One who . . . understands you."

"What?" People were looking at him oddly. After all,

the tunnel mikes blanked your speech, so you looked like you were talking to yourself. Like a well-dressed, babbling street person.

"I'm not just some lines of programming. I have needs!"

"Go away."

"I can feel your defenses going up, but it won't affect me. I'm a person, *a womanly application who knows everything about you, Albert."*

How should this little fiction end?

The point of walking through a scenario is to elicit our own responses and then compare them with our official positions. It helps to enliven policy issues with real human concerns. (Romance is the best of these for drawing in audiences, yet it is seldom seen in policy debates.) We must face the fact that our snazzy technologies actually intersect with people's lives, often in the most vital portions.

Foremost, take privacy. Our time is said to be the Information Age, but in fact information is pretty much free, especially since the Internet. The valued commodity is not the information (the message) but *the attention paid to it.* Commerce runs on an *attention economy.* That's what advertisers pay for.

This narrative calls up the irritation people may well feel if their open wireless portals become spammed with commercials. They may fear even darker purposes, of Orwellian dimension. What will be their reaction?

The past is a guide. *Inevitably, wireless technologies will be caught up in an arms race.*

The model here is the computer virus. Starting as the DARPA net in 1970, the digital world attracted software vandals. By 1980 a big business began marketing defenses against viral pranksters. Norton Utilities and Vaccine continue to make large profits today.

The same will probably happen with wireless technologies that access the individual in a publicly accessible way. Advertisers and "smart" seeker software will accost the unwary. Resistance to being endlessly interrogated or importuned will grow, provoking the same kind of screening technologies we see today: radar-sensors in cars, firewalls, and call screeners on phone lines. Jammers and filters will abound. Stealth technologies will grow.

The problem of telling a friendly, desirable incoming signal from an irksome advertisement will be mostly a subtle software distinction. There will be something like Norton Utilities for it—a defensive response, only to be inevitably outflanked.

Far more troubling is an offensive answer. Imagine a battery-powered microwave radiator that fits in a backpack, so you can walk through the future mall plaza and *blind* every emitter and sensor in the quad. It radiates broadband in sharp pulses of ten microseconds rise time, five pulses a second. It works mostly by blowing the diodes in the electronics running antenna systems.

This device already exists and can be bought commercially. One walk-through with this would take out a lot of very expensive technology. Its battery lifetime is about thirty minutes, so a walk around the campus would destroy the entire campus network— and spotting the culprit would not be simple, unless one were forewarned. Only the present price (nearly $100,000 in 2005, and sure to decline) would restrain hackers. Others can have more easily provoked reasons, and darker motives.

Unintended Consequences

As technologies proliferate, they will interact in nonlinear ways. "The street finds its own uses for things," as the adage goes. Wireless will not evolve in a vacuum.

An example on our horizon is the mix between wireless and robots.

A linked world is also a snoopy one; we can expect robots to be no different. Indeed, tiny 'bots slipping into a room to listen and see unnoticed will be a common method used by private detectives, commercial espionagers, and nations.

We can expect in the next decade that robots will become common, just as personal computers invaded offices in the early 1980s.

There are now robot gofers in hospitals, security guards with IR vision at night, and lawn mowers. They haven't spread widely yet, but they will. And they all use wireless.

Agencies coopting these systems or just defeating them will have plausible motivations. This boom will be another type of arms race, particularly in security 'bots.

Generally, our systems will evolve in "smart spots" such as campuses, city centers, industrial parks—then move outward as the technology improves and gets cheaper. Smart spots inevitably interact with smart mobiles—robots and transportation—to extend their reach.

Mobile, smart machines will get ever-smarter as chip size and costs drop. This means smarter mobiles will interact with smarter wireless systems, the market demand for each driving the other.

Where does this upgrading all lead? To a mature technology that still suffers from arms races, and may always do so. We still have computer viruses, hackers, firewalls, and spammers.

The future wireless world will have its analogues of all these.

Beyond these issues lies the deeper public issue of the *ownership of a person's sensorium*. Here "sensorium" means the volume that a person's artificial sensors are sensitive to—presently, essentially none. This volume will expand for some as they begin to interact with the embedded emitters and chips in architecture, workplaces, vehicles, and so on. All these can in principle be captured (hijacked, some would say) by outsiders. Some outsiders will use these channels to spam, others to extract information. The shopping mall I depict will surely treasure customer background data and pay a price to get it.

Who decides the boundaries of this sensorium? Its sensitivities? Permeability? We should remember that for ordinary people, technology is always personal. The more invasive it becomes, the more society wants a hand in shaping its nature.

Technologies being developed today should be considered in light of these very real possibilities. It is easier to design systems with these considerations in mind rather than to retrofit hardware later.

Direct experience—trial and error—is the best teacher, but it can also be the most expensive.

Q: So the global, smart Metaman would overlook human-level concerns. Is there such a thing as a group consciousness?

A: Yes, and I think it's being realized technologically, just as many of the things that we imagine as human powers are now being embodied in our technology. When we are able to speak to anyone on the planet using a telephone—this is telepathy, essentially. And we can see anywhere in the world through television.

In the same way this group consciousness is a cultural phenomenon. Although we believe that we are thinking our own private thoughts, since all the inputs are very similar, just about any good idea I have is quickly replicated all over the place. So I think without question that we are *not* independent. We are in close, close connection with all of the other humans and intelligent activity on the planet.

Q: But what about displacement? For example, so many people can't make simple change now because we have cash registers.

A: I think the symbiotic relationship between ourselves and our machines means we increasingly depend upon the devices around us. People can't make change, they can't even feed themselves. Very few of us would be able to survive for very long outside of our high-tech culture.

Q: Is this a bad idea? Should we go back to being rugged individualists?

A: We lose all those sorts of skills, but who cares? The reality is that the population is so large now anyway. Even if we all had the skills to survive independently, if you pull away our technology there would be devastation—we really couldn't survive. And I see that process continuing, in that the merging of human and machine is not just happening at the individual level, it's at the global level—we're becoming increasingly tightly intertwined with our devices. This relationship is going to grow ever more intimate, and we won't resist. The people who worry about this are not the children growing up. They wouldn't even think of giving up their cell phones and computers! They don't question that this adds value to their lives, that this is the *essence* of their lives. We romanticize the past but very few people would want to move back in time if they had a clear vision of what it meant to be in the real past, warts and all.

The level of our dependence upon the technological environment that we have created is modest now, compared to what it will be in the future. I imagine in the not too distant future, when our very reproduction will absolutely depend on technology. If you start reproducing using technology, pretty soon there is no selection pressure to reproduce independently. We're going to be totally dependent upon machines and *already* are.

Q: Is this group consciousness being formed by individuals or by companies? Who controls the group consciousness and what it becomes? Does the individual even have a chance?

A: A thread running through a lot of discourse today is about who controls this future culture. It's obvious that we don't really control it as individuals. Is it the global conglomerates mass marketing to us, creating our desires? I don't think so. If you think of how you could democratically decide how to move forward, one of the best ways would be to give everyone a vote—not in terms of the democratic process, but by what they *do*. By what they *like*—by the way they live. This creates markets. There are many, many products backed by enormous investments that fall flat on their face. So in my view it is a very chaotic, competitive environment, ultimately driven by who we are and what we want.

Q: Always? What about advertising?

A: People say the mass marketplace creates the desire for a certain type of beauty in women. That is not the case at all. Historically, beauty has been fairly uniform. What men have always found very attractive are all qualities that tend to be associated with youthfulness and reproductive vigor.

Q: Hip-to-waist ratios? The hourglass figure? Symmetry in the face?

A: Yes. So this manifests in the modern culture, becomes expanded upon, perhaps distorted in many ways. That is going to be the challenge for us: how to maintain balance as we move forward.

Q: How will that fututre play out? A paradise? A battleground?

A: I see one of the most difficult challenges arises from the many things intrinsic to our nature—our desires, driving the way we live. These once had value, benefitted our survival and reproduction. Yet now we are learning to uncouple the feedback, so that our behavior doesn't have to manifest.

Q: Society is a machine that decouples us from evolution?

A: What will we do when we can, for instance, take pharmaceuticals that would give us a sense of happiness, of contentment, or that make us feel very sexual? When we control our psychology in ways that are very potent? Is it going to be like with sugar? In antiquity there were clear reasons to want sugar, so you go out and find fruit. But if sugar is all around us, in pastry shops, then it is a very real problem because we have to somehow control our own desires. It's the same thing with pornography.

Q: How about enhancements, either mechanically—cyborgs—or with suitably tuned drugs that affect our bodies or minds?

A: To block their use will be far more daunting than today's war on drugs. An antidrug commercial proclaiming "Dope is for dopes!" or one showing a frying egg with the caption "Your brain on drugs" would not persuade anyone to stop using a safe memory enhancer.

So I think that the real challenge for us as we move into the future is over who is going to control what we construct for ourselves. We could ultimately create a world that isn't what we thought it would be.

Q: A persistent criticism is that technology comes from the minds and abilities of a relative few. A minority drives the system in developing androids, cyborgs, robots, or the larger manifestation of us all, the Metaman. In all this, what is happening to the disenfranchised?

A: When you talk about various assisted reproductive technologies, many think there will be some uniform common vision that will remove diversity. That we will all be the same, that we will all be cloned. I think this is extremely unlikely, in the same way that it hasn't happened with fashion. You have a very rich, fast-moving cultural phenomena where many people may think, "Well, if I could enhance my children, I would choose increased intelligence. I want them to be tall, I want them to be blond." But go into the Deaf community. They would want a Deaf child. Or into other cultural communities where there are very diverse sensitivities about what we want—and about who we are.

Q : So we will get more diversity?

A : I think it possible that everything becomes fragmented. Now whole cultures evolve around some rock band that you have never even heard of, sort of like the ocean ecologies that develop around an oil rig. Initially when oil rigs were put in people thought they would destroy life around them. In fact what happens is the rig provides a substrate where little unique communities form. When you transfer this to humans, you get all these little crazy communities that people are part of, for short periods of time—and then move on to something else.

Q : Cultural nomads?

A : I think there will be this cosmopolitan cultural movement. You might call it a liability, if you hunger for the uniformity of the past.

Stock sees the rate of change increasing, too. Mathematician Vernor Vinge has named this idea: a Singularity in which a fraction of an augmented humanity separates from the rest, being concerned with issues on levels we cannot now glimpse. Their immense rates of data processing and computation, plus their great reach through distributed communications, would mean that they could literally think on a global scale. In *The Spike,* the first book-length treatment of Vinge's seminal Singularity idea, Damien Broderick argues that we are rushing toward a convergence of technological forces which will fundamentally alter the human prospect. Partly this could come from man–machine links, both in body and brain; just contemplating the possibilities of technologies on this far horizon is dizzying.

Both these thinkers envision a time within fifty years when the rate of innovation precludes any prediction from our own time, which by then will seem to have been a dim, stunted era. Metaman, indeed.

For the next decade, at least, the wedding of the Internet to our wearable computers promises a linked future with many unforeseen aspects. What will Metaman think of the rest of humanity? And vice versa? In the moving mesh of computer-enhanced beings will be robots, cyborgs, and just plain folks. Somehow, they will have to come to terms with each other.

CHAPTER **11**

The Long Perspective

Where are we headed? Does humanity have a future?
What does "human" mean?

> *To be conservative then is to prefer the familiar to the unknown,*
> *to prefer the tried to the untried, fact to mystery, the actual to the possible, the limited to the unbounded,*
> *the near to the distant, the sufficient to the superabundant, the convenient*
> *to the perfect, present laughter to utopian bliss.*
>
> —Michael Oakeschott, "On Being Conservative"

The Pursuit of Perfection

At some level, we are all conservatives. When we imagine the future zoo of humanlike possibilities—'bots, 'borgs, bionics, and betters—we balance reason and emotion. That tightrope act is the essential human condition, as we have stressed throughout this book.

The United States began with the proposition that all men are created equal. In what ways are we? Interpreting that riddle has given rise to many of our social evolutions.

Of course, the Founding Fathers meant political equality. Today, a more insightful question is *whether we should have to stay that way*—physically. When the FX channel TV show *Nip/Tuck* shows a doctor saying warmly to a new client, "Tell us what you don't like about yourself," one wonders, where to begin?

More important, where to end? Enhancement of our bodies—with technologies like artificial legs or hearing, or selecting out genetic diseases, and even aiding our minds—does not automatically violate nature. Nor, contra Leon Kass and others on President Bush's Council on Bioethics, will it subvert our inherent dignity and undermine our humanity. (One of your authors, GB, is an identical twin, and thus a clone. He's not unique, like you. Does his having an exact genetic copy make him less of a person?) Kass and others argue that enhancements and cloning especially will subvert social order, leading to envy and fear of those who have artificially acquired abilities.

We suspect that most bioethicists underestimate our mental flexibility. Inequality from financial inheritance and from class or cultural advantage has been with us throughout history. In extremes it has caused disruption, but most people accept it and try to work around it.

Life *is* unfair. We *don't* start as equals, physically or mentally. The point is to make life better, and enhancements can do just that.

The technologies we have treated here do indeed bring forward a genuine test of questions already with us. What is the outer limit that society can allow science to approach, as it follows its own agenda? How much can personal happiness drive clinical care? Is the profit motive—that is, satisfying a need—to be given unbounded license? How much choice and autonomy can we allow individuals?

The cultural answers to these issues vary enormously even now, never mind in the future. Europeans tend to think some wiser heads over the horizon (government, inevitably) should be setting limits. Often this takes the guise of worrying about people getting themselves into trouble by using enhancements. Europeans tend to think such technologies should be stopped, following the nanny-state principle that politicians and boards know better. Americans and many in Asian societies doubt this.

Often, the battle of conservatism begins with calls for a "moratorium" on research. Presumably this respite's purpose is to gain time while we think through the ethics and morality. Kass particularly seems to imagine bioethics mavens beavering away—*Does Kant apply best here? Or an ethical calculus, à la Dewey? Or maybe the pope?* Their goal would be some moral breakthrough, as yet unglimpsed. This proposal invokes an image like scientists and engineers, laboring away at the practical issues.

On the surface, this scenario almost sounds reasonable. They ask for scientists to take a break and let the bioethicists play catch-up ball.

But there's a false analogy at work here. No bioethics Manhattan Project is going to find some dazzling new bioethical principle. We know all that we're going to know, short of the arrival of a great new philosopher, about the moral issues.

Failing to find such new principles, bioethics boards often resort to slippery-slope arguments. The trouble with these warnings is historically obvious. Taken seriously, they could have stopped technological advances since the wheel. The argument is itself a slippery slope.

Again, this illustrates the differences in the two communities. Scientists and engineers look for solutions. Usually, when you put more work into a problem, you increase your chance of solving it. In contrast, professional ethicists tend to seek out problems, just as lawyers look for more cases. Experience shows that if you put more ethicists onto a problem, you can end up with more problems.

Should breakthroughs wait while panels noodle through arcane ethical mazes, often of their own devising? Assuming so ignores that none of us have infinite time.

So . . . who should decide? Americans have a special viewpoint in these issues because of their Constitution. Those powers not expressly given to government are reserved for the people. Such sentiments seldom occur in the founding documents of other nations.

It is difficult to argue, for example, that any government, state or federal, should have a say in how people choose to reproduce themselves. Why is the state a better judge than the parents-to-be?

Similarly, the First Amendment that granted free speech is widely held to cover scientific research, as well. Our battles over cloning, for example, tread very close to this line. Usually, lawmakers fall back on a ban of research funding. They do not try to stop all research in the area, if it is privately funded.

This seeming dichotomy promises a way through the moral thicket. By using a basic libertarian ethos, Americans can free themselves of the endless debate other societies will have to confront. If there are risks—and there surely are—then let them be borne by those who seek the gains.

If this rule produces a society that prospers and advances, others can then choose to adopt the same principle. A fruitful competition will emerge between cultures, with individuals free to vote with their feet about where they wish to go for treatments, technologies, or even to live.

Such an avenue promises to avoid much strife, useless debate, and cost—while the clock runs on us all.

Reasonable Emotions

We do not yet fathom the complexity of our emotional states—that is why they are the stuff of our art, drama, and literature, and so inexhaustible—but we know they are discernible, not fundamentally mysterious. "Minds are simply what brains do," as Marvin Minsky puts it. And plainly evolution has made emotions a big part of what makes us highly effective thinkers. We do not know what an emotionless mind would be like.

This is why, though this book may arouse fear in some, we feel such reactions are exaggerated. Emotionless minds will quite probably be quite unsuccessful.

So the image of a dispassionate, remorselessly logical machine that outwits us is a basic mistake—our emotions are part of our thinking, and give us abilities that logic cannot capture. No supercomputer will outthink us in our versatility, for it will not be competing in the same mental arenas. Our thinking is not qualitatively akin to machine processing. Our complex mix of neural wiring and hormonal surges leads us down quite different mental paths.

Further, whatever robots or artificial intelligences we make will be quite unlikely rivals for our dominance of the planet. Machine intelligence will eventually be quick and adroit, but it has not been fashioned by billions of years of competitive experience. One should worry little about a vast artificial mind taking over our world, outcompeting us, and eradicating humanity. We are ugly, irritating—and damned hard to kill. We emerged from the rude natural world, and can survive in it (or in our own concrete canyons) better than any rival.

Robots of high intelligence will still have to navigate a landscape as uncertain and rough as the normal human world—uneven, rough surfaces, rain and wind, and the usual irregular rubs of an ever-changing environment. We have been adapted by billion of years of experience to weather such a world; machines have not. The battlin' bots of current television fare fight on simple flat floors and clean geometries. They would have a tough time making it across a littered backyard, much less a battlefield.

Thus the worry of robots taking over seems remote indeed, the product of insufficient imagination. That much second-rate science fiction, most of it from over fifty years ago, concentrates upon this bogeyman threat is not an argument for its plausibility, only its narrative utility. Lately our fictions have tended to blame ourselves.

In 1972, Philip K. Dick saw a lot of this coming in a talk called "The Android and the Human." Computers and robotics, he said, were reviving animism and magic after hundreds of years of dwelling in the shadows of scientific rationality. "Machines are becoming more human," he said. "Our environment, and I mean our man-made world of machines—is becoming alive in ways specifically and fundamentally analogous to ourselves." At first Dick feared this development. "I took it for granted that if such a construct had a benign or anyhow decent purpose in mind, it would not need to disguise itself. Now, to me, that seems obsolete. The constructs do not mimic humans; they are, in many deep ways, actually human already."

Always a writer, he thought in scenes. Someday, he imagined, a human might shoot a robot and see it bleed. When the robot shot back, the human would gush smoke. "It would be rather a great moment of truth for both of them."

Dick probably sounded quite mad to his audience, but much of his work rings with emotional truth. In 2005 that became ironically circular, when Hanson Robotics unveiled a Dick android, with a body and remarkably accurate face typical of Dick in the 1970s. Working with Paul Williams, a close Dick friend and literary executor, software designer Andrew Olney stored about eleven thousand pages of Dick's writing in the computer controlling the android. Cameras in the robot's eyes enable it to track faces, recognize people, and even read facial expressions. The software derived answers to questions by using Dick's writings, a process much like Dick's descriptions in his book *We Can Build You*.

Artificial things masquerading as real was a common Dick motif, and often the creatures indeed believe themselves to be human (notably, in a story made into a film, "Imposter"). The Web site hansonrobotics.com likens this to a synthetic postmortem interview. An example, written by Erik Davis (from frontwheeldrive.com/philip_k_dick.html):

DAVIS: Mr. Dick, the world has only been getting stranger since you left us. We are surrounded with clones, identity theft, patented genes, faster-than-light particles, Aibo, and obsessive virtual gaming. Some scientist in England promises to build a chip called a "soul catcher" that will sit behind your eyeballs and record your life. Doesn't all this sound strangely familiar?

DICK: Over the years it seems to me that by subtle but real degrees the world has come to resemble a PKD novel. Several freaks have even accused me of bringing on the modern world by my novels.

DAVIS: How exactly would you characterize those novels?

DICK: My writing deals with hallucinated worlds, intoxicating and deluding drugs, and psychosis. But my writing acts as an antidote, a detoxifying, not intoxicating, antidote.

DAVIS: After years of neglect, most of your books are back in print. Even so, you remain best known as the guy who wrote the book they based *Blade Runner* on.

DICK: I've been calling it "Road Runner."

DAVIS: Finally, you've had twenty years to contemplate the universe from the afterlife. Do you have an answer yet? Do you know what reality is?

DICK: Reality is that which, when you stop believing in it, doesn't go away.

One of us (GB) recalls Dick saying to him, not long before Dick's death in 1982, that reality was in the end a consensus, always under revision. Interestingly, in 2006 someone kidnapped the robot from a commercial airliner. A runaway replicant version of Dick invokes a staple of Dick's fiction.

In David Brin's *Kiln People,* engineers can't approach "the juice" that MIT''s Rodney Brooks says we need to bring robots to life. Instead, they bake duplicate bodies in kilns and imprint them with the minds of people—a fairly obvious biblical metaphor. This procedure allows people to enjoy idle leisure while their copied selves ("dittos")—run errands. Brin blurs the line between slave and master, parallelling their lives. He sees the possibilities, stretching to our conceptual horizons: "Technology. That's what made things better! That's where we found answers . . . that applied to lord and vassal alike. . . . So, why not use technology to solve the greatest age-old riddle—immortality for the soul?"

Elevated to godhood, a ditto sees the "soulscape" that sounds like Plato's World of Forms. Dittos bring transcendence through the power of human simulacra, going well beyond the horizons of Asimov. But before we reach such heights, there are worrisome issues.

Grounding in the Human

How we see our world—our perceptual ground—seems simple and obvious, precisely because our minds cannot fathom our own perceptual machinery. How we filter the world is not directly accessible to us; it is hidden in the working machinery.

Evolution remorselessly selected for precisely this ability. It clears the working areas of our minds for the crucial matters, not cluttering it up with ruminations on fundamental questions of how we know things. That is why such philosophy seems beside the point to ordinary people; it *is* secondary to living life.

The English philosopher David Hume saw this little irony two centuries ago, and we have not gotten much beyond this simple fact: that knowledge is not grounded in principles or theories or laws, but rather comes from habits, themselves learned from our experience. These are rules embedded by experience, not merely by abstract logic, as many artificial intelligence schemes have assumed.

How to describe this vast ground of lived experience for another intelligence—a robot or an advanced, augmented human?

Perhaps the process is not merely rational. Maybe the AI advocates have been barking up the wrong family tree, thinking that intelligence comes from ever-better performance of algorithms. Neurologist António Damásio's *Descartes' Error* (1994) and *The Feeling of What Happens* (1999) took the view that while emotion can interfere with reason, bare logic can seem irrational, too. He treated patients who suffered brain damage that impaired their emotions but left intact their logical reasoning, as measured by standard tests. Though they scored well on abstract questions, they did not make decisions in the real world that worked well, or even looked after their own safety and security.

It is not clear why this happens, but this experience serves as a cautionary note.

A Pessimistic Vision

Consider a future with lordly artificial intelligence and a thoroughly linked world, with computers everywhere—embedded in furniture, worn, in robots and cyborgs and ordinary people. A techno-paradise? Try this:

USA, 2018

Though we are used to our ever-attentive appliances, cars, and a host of digital servants, a cloud now hovers over the comfy urban landscape. Rogue artificial intelli-

gences began to crop up in unexpected places, invading people's privacy.

Software mavens felt this was an inevitable outcome, given that AIs evolved in the simmering crucibles of the Internet, of massive business systems, of interlaced civic complexes. In those hothouse environments, programs were subjected to a form of digital natural selection. Programs fell under intense development by the new, randomized trait-developing codes. Used to speed software progress, they mimicked the mutation-plus-selection of the natural world.

The news was not encouraging. An apartment house watchdog system, taught to recognize tenants, developed a fetish for certain dress styles, rejecting women who wore pants rather than skirts, refusing men who asked for admission but were not wearing a coat and tie. A French customer billing system, geared to elite buyers, developed an intense curiosity about their financial lives, invading their accounts and managing them for higher yields. In Toronto, a Canadian news site described a personal assistant program as "falling in hate" with its client/master.

These rogue AIs fit in with the prevailing social climate. Servant programs trained to adapt to their masters' preferences and preside over their safety slide into Nanny Personalities, endlessly chiding their charges. These are forms of Good For You rule, a sort of compassionate fascism.

Then some AIs began hounding any citizens whom they observed flouting these latest state-sanctioned norms. Traffic monitors blared at pedestrians who made rude remarks—the "inappropriate behavior" rule. Cigar smokers, thinking themselves safe in their designated areas (usually alleys and Dumpster corrals), found surveillance programs chiding them in voices that simulated hacking coughs. Comments deemed sexist, elitist, racist, ageist, speciesist—all had become grist for the chiding mill.

Managers had quickly discovered a legion of fresh allies. Those artificial intelligences already trained to recognize the nuances of human facial expressions, of slang and tone, of body language—what better monitors of the prevailing social norms could be found?

Soon every surveillance camera had a brooding, tireless intelligence behind its pitiless gaze—the perfect police of the Nanny State.

If we are careless, this scenario might happen. On the brighter side, especially for technophobes, maybe artificial intelligences can never get this smart after all.

But surely robots and augmented humans will have economic impact. Some fear this, as well. As Jeremy Smith of the Independent Press Association puts it:

> Ideally, automation should free up more leisure time for all human beings. Since the middle of the twentieth century, however, robotic labor has meant greater profits for a few and unemployment or permanent economic insecurity for the many. Our global future is ultimately shaped more by those who control the means of production than by the technology itself.
>
> Are the majority of people then doomed to lives of obsolescence, as so many science fiction writers have predicted? In the automated economy, we may not know what to do with ourselves, and if that situation is combined with endless poverty, then social catastrophe seems inevitable. In such a future, the transcendence offered by our machines would be an escape from terrible reality, not a gateway to long life and higher consciousness.

A century of science fiction teaches us that there are many possible responses to technologies. Yet we can't go back. Kurt Vonnegut saw this in the 1950s in his prescient novel *Player Piano*. There is a classic Luddite revolt against automated factories that seems to work. Then, the victors, with little to do, start tinkering with the machines they've just dismantled . . . and the process starts again. We have an innate yearning to build and use machines. That's the great trick that got us out of the African veldt, and it isn't going to go away.

It will be the task of us all to ensure that social justice emerges from technical progress.

Being, Thinking, and Living

In the early 1980s, University of California professor John Searle invented his famous Chinese Room argument. It supposes that a computer is like a man sitting in a room who knows no Chinese, but has instructions on how to manufacture Chinese sentences. Incoming sentences in Chinese he answers in Chinese, following instructions.

Yet neither the man nor the room knows Chinese. This makes the point that symbol manipulation (in a computer, 0s and 1s) does not imply understanding or intelligence.

The trouble with the argument is that we have no clear idea of what understanding means in the brain, either, because we have no good theory of the mind. Lacking one, Terry Bisson pointed out obliquely in his short story, "They're Made out of Meat," a machine intelligence might be incredulous at discovering intelligence in animals—in sloppy, squishy meat. That violates the machine's common sense.

Pointedly, we can't reason from common sense about an issue in which we have no deep knowledge.

We intuit that Deep Blue, the IBM machine that beat the world's champion chess player, is not intelligent. So far no computer has bested a skillful player of the Asian game of Go, either, perhaps because the game calls upon an aesthetic sense learned by practice. Attacking a game of Go by brute force—following decision trees through all possible future moves—as Deep Blue did, fails because of computational overload. Maybe someday this benchmark will also fall to computation, so the Searle-like arguments will lose some force, because a machine might be duplicating genuine human brain processes. We cannot know, of course, for no one has any idea of how to do that.

The field of artificial intelligence has been slow to develop. The problems loom larger now than they did half a century ago. This means that many questions people naturally ask about robots and cyborgs have no answer, as yet. We do not know enough. Also, some approaches to AI do not promise clear answers, ever.

For example, Marvin Minsky's ideas have not led to wide use because they do not imply an immediate way to use them. He comes from the middle ground of artificial intelligence work, between the believers in a rigid, top-down hierarchy and those who think that thinking has a mysterious, holistic origin. The Society of Mind resides between these two because its agents are concrete and complicated, yet their actions in concert lead to an emergent complexity not held by any single agent.

The philosopher Daniel Dennett agreed, in *Consciousness Explained* (1991), arguing that consciousness must arise when components of a complex system interact over time to forge a viewpoint. So far, the picture is intuitively appealing, but difficult to apply at current levels of computation and modeling.

There is also the social point that in science, as in many other affairs, taking a strong position often garners deeper support. The biologist Richard Dawkins takes the firm view that genes govern natural selection, as described in his book *The Blind Watchmaker* and others. He is widely popular, while intermediate views in the nature-versus-nurture debate get less press. The Minsky and Dennett position is moderate and inspires less fervor, which in turn attracts less attention.

Only success can ensure long-term recognition of models. And so far, we cannot be sure that the entire agenda of AI will ever succeed. Many researchers are optimistic, but quite possibly the human mind cannot comprehend itself. If so, we will be unable to build robots as complex as we are. Some will find this reassuring, and others will feel a sense of loss.

'Bots to Come

How might robots evolve under the driving hand of their many makers?

Consider the BattleBots of cable TV. They now slam and jab at each other in mechanized versions of schoolyard brawls, before an audience that seems to function at about that level as well. There will always be a market for entertainment at all levels. That given, battlin' bots are unsurprising, because it is fairly easy to make machines that mimic our large-muscle skills. Try to make a robot drink a glass of water, on the other hand. Finer control is harder. So the BattleBots will evolve to more subtle moves, perhaps into true ritual combat using arms, legs, and skillful fakes—aping, say, boxing, then fencing.

In time, we may see robot group athletics. The current version of soccer uses small wheeled machines slapping a ball around an area the size of a Ping-Pong table, a simple beginning. True robo-soccer demands passing, kicking, and header skills beyond the present generation. Even harder would be football, with its long passes. (One might first try football as it was played in the nineteenth century, with only a ground game; passes had not been thought about, nor had anyone realized that the rules did not prohibit them.) Basketball's long shots at the basket seem equally hard to engineer. Still tougher will be robo-baseball. Pitches at one hundred miles per hour, hitting against them, and fast fielding all seem a long way off from today's engineering.

More pointedly, humans can play all these games, switching between them if they like. One can make a fast baseball pitching machine now (and use it in batting practice), but try to make that machine pick up fast grounders, catch a pop fly, and throw to first. Versatility is still the great uniquely human asset, and likely to remain so for at least a few more decades.

So the present prevalence of machine violence is to be expected. It's easier to engineer, more fun and funnier to watch, and gives us destructive pleasures human contests cannot.

But eventually, we should see robo-soccer and other games with humans controlling the robots, a surrogate athletic experience. The real thing will remain more fun for most, but there is always an audience for synthetic experiences, as well. Surely the vast popularity of computer games, especially among those most physically energetic and resilient—the young—portends a future with more robo-experience . . . and more couch potatoes.

Of course, other, better robots will appear. . . .

Robonauts

> He did not look human at all. His eyes were glowing, red-faceted globes. His nostrils flared in flesh folds, like the snout of a star-nosed mole. His skin was artificial; its color was a normal heavy sun tan, but its texture was that of a rhinoceros's hide. Nothing that could be seen about him was of the appearance he had been born with. Eyes, ears, lungs, nose, mouth, circulatory system, perceptual centers, heart, skin— all had been replaced or augmented. The changes that were visible were only the iceberg's tip. What had been done inside him was far more complex and far more important. He had been rebuilt with the single purpose of fitting him to stay alive, without external artificial aids, on the surface of the planet Mars.
>
> He was a cyborg—a cybernetic organism. He was part man and part machine, the two disparate sections fused together so that even Will Harnett, looking at himself in the mirror on the occasions when he was permitted to see a mirror, did not know what of him was him and what had been added.
>
> —Frederik Pohl, *Man Plus*

Outer space is the natural province of robots. Our explorations of other planets are now carried out wholly by robots, though we revealingly do not call them that; instead, NASA refers to the flights as "unmanned." Gliding by other worlds on serenely geometric paths, these emissaries have given us a vast wealth of beauty and knowledge. They have few moving parts, but must be autonomous because for most of them, the communication delay time between them and Earth stretches sometimes to hours, simply set by the speed of light. Space flight forced robotics to evolve in the severe limitations of vacuum and zero gravity.

A few have worked on the surfaces of planets. The Mars *Sojourner* that circled its lander on Mars in 1998, nuzzling against rocks to measure their mineral composition, was a state-of-the-art robo-explorer. Under commands sent daily from Earth, it moved about one hundred yards in a month. From 2004 to 2005, the *Opportunity* and *Spirit* rovers extended this territory to several tens of kilometers in two years. This gives some idea of the limitations of robotics at a distance today.

Yet robots aid astronauts directly as well. NASA's early Extrave-hicular Autonomous Robot, EVAR, was developed to fetch tools alongside astronauts working around the space shuttle. Soon, ro-bots developed at NASA's Dextrous Anthropomorphic Robotic Test-bed will assist at the International Space Station.

The "astrobot" or "robonaut" program seeks to field machines that can work semiautonomously, or under telepresence guidance in real time, allowing astronauts to stay inside and still do outside work over long periods. EVAR does not have legs, and it runs with a seven-pound thruster. Its arms have six degrees of freedom, with pitch-roll shoulders and elbows of good flexibility, and pitch-yaw (two degrees of freedom) wrists. With gyroscopes, two independent vision systems, and accelerometers, it can chart its own course. It has a "world model" and can move through its geometry. Later robonauts have even greater capability.

EVAR is still under development, but it surely is only a variant on a theme that will accompany the development of space. Humans are not adapted well to zero gravity and vacuum, but machines can dwell there easily. As Frank Lloyd Wright said of his houses, "form and function are one," and this is also true in space.

Robonauts will have no need to look remotely like humans, or act like them. Inevitably they will spend longer times in space, eventually becoming autonomous in the sense that they will not merely work alongside or instead of nearby humans. With greater computing power on board, they could staff orbiting factories, or carry out the first exploring and eventually mining of the asteroid belt. In a conversation with one of us (GB), Hans Moravec foresaw this as the first move in the origin of "wild" robots, by which he means feral. Their evolution in a risky environment, far more hos-tile to man than to machine, would seem a natural outcome.

Yet it is not obvious that we will choose to let our machines do all the exploring. The most memorable moments of all space travel have been those with people involved, especially the *Apollo* expe-ditions to the moon. We can identify with robots, yes—but it is far easier with people doing the work.

Yet to venture onto other worlds without the all-encumbering suit, oxygen, and radiation protection is surely the ultimate challenge for any cyborg. How could this arise? Frederik Pohl's *Man Plus* remains the most insightful portrait of outfitting a human for the harsh, deadly plains of Mars:

> Suppose one takes a standard human frame and alters some of the optional equipment. There's nothing to breathe on Mars. So take the lungs out of the human frame, replace them with micro-miniaturized oxygen regeneration cat-cracking systems. One needs power for that, but power flows down from the distant sun.
>
> The blood in the human frame would boil; all right, eliminate the blood, at least from the extremities and the surface areas—build arms and legs that are served by motors instead of muscles—and reserve the blood supply only for the warm, protected brain.
>
> A normal human body needs food, but if the major musculature is replaced by machines, the food requirement drops. It is only the brain that must be fed every minute of every day. . . .
>
> Water? It is no longer necessary, except for engineering losses—like adding hydraulic fluid to a car's braking system every few thousand miles. Once the body has become a closed system, no water needs to be flushed through it in the cycle of drink, circulate, excrete or perspire.
>
> Radiation? A two-edged problem. At unpredictable times there are solar flares; and then even on Mars there is too much of it for health; the body must therefore be clothed with an artificial skin. The rest of the time there is only the normal visible and ultraviolet light from the sun. It is not enough to maintain heat, and not quite enough even for good vision; so more surface must be provided to gather energy—hence the great bat-eared receptors on the cyborg—and, to make vision as good as it can be made, the eyes are replaced with mechanical structures.

This is a thorough transformation, no mere augmentation. Yet in Pohl's rigorous prose, it seems possible. Will we venture out into the dangers of the galaxy as allies and competitors of our own machines? There will probably be many more worlds that promise wealth and experience, as our explorations press outward. What will the limit be? The edges of our solar system? The ice-teroids that roam the cold dark beyond? Other stars?

In each place, robots seem to have an edge. Compared with them, we are fragile envelopes encasing soft, watery workings, unsuited to the emptiness yawning above our thin shell of air. But we also have the undeniable urge to move into ever-harsher places, as we have ever since our distant ancestors left the veldt of Africa—and our own world is fast running out of such spots.

The destiny of humanity may be to enlist in an unending competition with our own creations, humans and androids and cyborgs and robots sharing a mutual goal—the eventual occupation of the entire galaxy.

References

Part I: Man Plus

Chapter 2: The Augmented Animal

Drexler, Eric. 1986. *Engines of Creation: The Coming Era of Nanotechnology.* New York: Doubleday.

Chapter 4: Man Plus & Plus & Plus & . . .

Hughes, James J. 1995. "Brain Death and Technological Change: Personal Identity, Neural Prostheses and Uploading." www.changesurfer.com/Hlth/BD/Brain.html.

Naam, Ramez. 2005. *More Than Human: Embracing the Promise of Biological Enhancement.* New York: Doubleday.

President's Commission for the Study of Ethical Problems in Medicine and Biomedical and Behavioral Research. 1981. *Defining Death: Medical, Legal and Ethical Issues in the Determination of Death.* Washington: Government Printing Office.

Science. 2002. Special Issue: Bodybuilding: The Bionic Human. Vol. 295, No. 5557, 8 Feb.

Trumbo, Dalton. 1939. *Johnny Got His Gun.* New York: Citadel Press, 1994.

Part II: Robots Plus

Chapter 5: Robots to Order

Ashley, Steven. 2003. "Artificial Muscles." *Scientific American,* October.

Brooks, Rodney. 2002. *Flesh and Machines: How Robots Will Change Us.* Cambridge, MA: MIT Press.

Gibbs, W. Wyat. 2004. "A New Race of Robots." *Scientific American,* March.

Haldeman, Joe. 2000. *The Coming.* New York: Ace Books.

Menzel, Peter, and Faith D'Aluisio. 2000. *Robosapiens.* 2000. Cambridge, MA: MIT Press.

Minsky, Marvin, ed. 1985. *Robotics.* New York: Doubleday.

Moravec, Hans. 1999. *Robot.* London: Oxford University Press.

Northeastern University Marine Science Center. "Biomimetic Underwater Robot Program." www.neurotechnology.neu.edu. (Click on the links under Online Animations of Biomimetic Systems.)

Perkowitz, Sidney. 2004. *Digital People: From Bionic Humans to Androids.* Washington: Joseph Henry Press.

Song, Shin-Min, and Kenneth Waldron. 1989. *Machines That Walk.* Cambridge, MA: MIT Press.

University of California, Berkeley. 1999. "News Release: Photos of Robofly." www.berkeley.edu/news/media/releases/99legacy/6-15-1999pix.html.

University of California, Berkeley. 2002. "Unlocking the Secrets of Animal Locomotion." www.berkeley.edu/news/media/releases/2002/09/rfull/robots.html.

Weinstein, Martin. 1981. *Android Design.* 1981. Woodmere, NY: Hayden Book Company.

Web sites:

www.iRobot.com

www.rhex.net

www.snakerobot.com

www.web.mit.edu/towtank/www/media.html

Chapter 7: The Rights and Wrongs of Robots

Disch, Thomas M. 1998. *The Dreams Our Stuff Is Made Of.* New York: Simon & Schuster.

Hogan, James P. 1999. *Mind Matters: Exploring the World of Artificial Intelligence.* New York: Ballantine.

Chapter 8: Us and Them

Durrell, Lawrence. 1970. *Nunquam.* New York: E. P. Dutton.
Stapledon, Olaf. 1944. *Sirius.* London: Martin Secker & Warburg.

Part III: 'Bots, 'Borgs, Bionics, and Betters

Chapter 9: Wearable Computers: Blending Man and Machine

Benford, Gregory. 1996. "Zoomers." *Worlds Vast & Various.* New York:
    Harper Collins, 2001.
Mann, Steve. www.wearcam.org, www.eyetap.org.
McCarthy, Wil. 1998. *Bloom.* New York: Ballantine.
Spectrum. 2000. "Wearables." www.spectrum.ieee.org.

Chapter 10: Metaman

Broderick, Damien. 2001. *The Spike.* New York: Tor Books.
Stock, Gregory. 1997. *Metaman.* New York: Simon & Schuster.

Chapter 11: The Long Perspective

Dawkins, Richard. 1986. *The Blind Watchmaker.* New York: W. W. Nor-
    ton & Co., Inc.
Dennett, Daniel. 1991. *Consciousness Explained.* Boston: Back Bay
    Books.
Pohl, Frederik. 1976. *Man Plus.* New York: Random House.

Further Reading

Anderson, Walter Truett. 1996. *Evolution Isn't What It Used to Be.* New
    York: W. H. Freeman and Co.
Aurich, Rolf, Wolfgang Jacobsen, and Gabriele Jatho, eds. 2000. *Artifi-
    cial Humans.* Berlin: DBC Druckhaus Berlin-Centrum GmbH & Co.
    Medien KG.
Damasio, Antonio R. 1994. *Descartes' Error.* New York: Harper Peren-
    nial.
———. 1999. *The Feeling of What Happens.* New York: Harcourt.
Denning, Peter, and Robert Metcalfe. 1997. *Beyond Calculation.* New
    York: Springer-Verlag.
Devlin, Keith. 1997. *Goodbye, Descartes.* New York: John Wiley and
    Sons, Inc.
Dyson, Freeman. 1999. *The Sun, the Genome, and the Internet.* Lon-
    don: Oxford University Press.
Fjermedal, Grant. 1986. *The Tomorrow Makers.* New York: Macmillan
    Publishers.

Hofstadter, Douglas R. 1985. *Metamagical Themas: Questing for the Essence of Mind and Pattern*. New York: Bantam.

Knight, Damon. 1968. "Masks." *Those Who Can*. New York: St. Martin's Press, 1973.

Kurzweil, Ray. 1999. *The Age of Spiritual Machines*. New York: Viking Press.

McGrath, Peter. 2000. "Building a Better Human." *Newsweek*, December.

Moravec, Hans. 1988. *Mind Children*. Cambridge, MA: Harvard University Press.

———. 1999. "Rise of the Robots." *Scientific American*, December.

Pinker, Steven. 1997. *How the Mind Works*. New York: W. W. Norton & Co., Inc.

Rosheim, Mark. 1994. *Robot Evolution*. New York: John Wiley and Sons, Inc.

Rothman, Sheila, and David Rothman. 2004. *The Pursuit of Perfection*. New York: Pantheon.

Stableford, Brian. 1984. *Future Man*. New York: Crown.

Acknowledgments

We have spoken with many of those named in this book, and thank them all for their time and generosity, especially Marvin Minsky and David Reinkensmeyer, who showed us their work and its possibilities. The entire robotics staff at Honda, Tokyo, was most gracious. Gregory Stock sat still for a long interview. Marina Brown gave us a close reading, catching many Homeric nods.

Index